ESI Lectures in Mathematics and Physics

The Erwin Schrödinger International Institute for Mathematical Physics
Boltzmanngasse 9
A-1090 Wien
Austria

The Erwin Schrödinger International Institute for Mathematical Phyiscs is a meeting place for leading experts in mathematical physics and mathematics, nurturing the development and exchange of ideas in the international community, particularly stimulating intellectual exchange between scientists from Eastern Europe and the rest of the world.

The purpose of the series *ESI Lectures in Mathematics and Physics* is to make selected texts arising from its research programme better known to a wider community and easily available to a worldwide audience. It publishes lecture notes on courses given by internationally renowned experts on highly active research topics. In order to make the series attractive to graduate students as well as researchers, special emphasis is given to concise and lively presentations with a level and focus appropriate to a student's background and at prices commensurate with a student's means.

Previously published in this series:
Arkady L. Onishchik, *Lectures on Real Semisimple Lie Algebras and Their Representations*

Werner Ballmann

Lectures on Kähler Manifolds

European Mathematical Society

Author:

Werner Ballmann
Mathematisches Institut
Rheinische Friedrich-Wilhelms-Universität Bonn
Beringstraße 1
D-53115 Bonn
Germany

2000 Mathematics Subject Classification: 53C55, 32Q15, 32Qxx, 58Jxx, 53C35

Bibliographic information published by Die Deutsche Bibliothek

Die Deutsche Bibliothek lists this publication in the Deutsche Nationalbibliographie;
detailed bibliographic data are available in the Internet at http://dnb.ddb.de.

ISBN-10 3-03719-025-6
ISBN-13 978-3-03719-025-8

Contact address:

European Mathematical Society Publishing House
Seminar for Applied Mathematics
ETH-Zentrum FLI C4
CH-8092 Zürich
Switzerland

Phone: +41 (0)44 632 34 36
Email: info@ems-ph.org
Homepage: www.ems-ph.org

Typeset using the author's TEX files: I. Zimmermann, Freiburg
Printed in Germany

9 8 7 6 5 4 3 2 1

To my wife Helga

Preface

These are notes of lectures on Kähler manifolds which I taught at the University of Bonn and, in reduced form, at the Erwin-Schrödinger Institute in Vienna. Besides giving a thorough introduction into Kähler geometry, my main aims were cohomology of Kähler manifolds, formality of Kähler manifolds after [DGMS], Calabi conjecture and some of its consequences, Gromov's Kähler hyperbolicity [Gr], and the Kodaira embedding theorem.

Let M be a complex manifold. A Riemannian metric on M is called Hermitian if it is compatible with the complex structure J of M,

$$\langle JX, JY\rangle = \langle X, Y\rangle.$$

Then the associated differential two-form ω defined by

$$\omega(X, Y) = \langle JX, Y\rangle$$

is called the Kähler form. It turns out that ω is closed if and only if J is parallel. Then M is called a Kähler manifold and the metric on M a Kähler metric. Kähler manifolds are modelled on complex Euclidean space. Except for the latter, the main example is complex projective space endowed with the Fubini–Study metric.

Let N be a complex submanifold of a Kähler manifold M. Since the restriction of the Riemannian metric of M to N is Hermitian and its Kähler form is the restriction of the Kähler form of M to N, N together with the induced Riemannian metric is a Kähler manifold as well. In particular, smooth complex projective varieties together with the Riemannian metric induced by the Fubini–Study metric are Kählerian. This explains the close connection of Kähler geometry with complex algebraic geometry.

I concentrate on the differential geometric side of Kähler geometry, except for a few remarks I do not say much about complex analysis and complex algebraic geometry. The contents of the notes is quite clear from the table below. Nevertheless, a few words seem to be in order. These concern mainly the prerequisites. I assume that the reader is familiar with basic concepts from differential geometry like vector bundles and connections, Riemannian and Hermitian metrics, curvature and holonomy. In analysis I assume the basic facts from the theory of elliptic partial differential operators, in particular regularity and Hodge theory. Good references for this are for example [LM, Section III.5] and [Wa, Chapter 6]. In Chapter 8, I discuss Gromov's Kähler hyperbolic spaces. Following the arguments in [Gr], the proof of the main result of this chapter is based on a somewhat generalized version of Atiyah's L^2-index theorem; for the version needed here, the best reference seems to be Chapter 13 in [Ro]. In Chapter 7, I discuss the proof of the Calabi conjecture. Without further reference I use Hölder spaces and Sobolev embedding theorems. This is standard material, and many textbooks on analysis provide these prerequisites. In addition, I need a result from the regularity theory of non-linear partial differential equations. For this, I refer to the lecture notes by Kazdan [Ka2] where the reader finds the necessary statements together with precise references for their proofs. I use some basic sheaf theory in the proof of the Kodaira embedding theorem in Chapter 9. What I need is again standard and can be found, for example, in [Hir, Section 1.2] or [Wa, Chapter 5]. For the convenience of the reader, I include appendices on characteristic classes, symmetric spaces, and differential operators.

The reader may miss historical comments. Although I spent quite some time on preparing my lectures and writing these notes, my ideas about the development of the field are still too vague for an adequate historical discussion.

Acknowledgments. I would like to acknowledge the hospitality of the Max-Planck-Institut für Mathematik in Bonn, the Erwin-Schrödinger-Institute in Vienna, and the Forschungsinstitut für Mathematik at the ETH in Zürich, who provided me with the opportunity to work undisturbed and undistracted on these lecture notes and other mathematical projects. I am grateful to Dmitri Anosov, Marc Burger, Thomas Delzant, Beno Eckmann, Stefan Hildebrandt, Friedrich Hirzebruch, Ursula Hamenstädt, Daniel Huybrechts, Hermann Karcher, Jerry Kazdan, Ingo Lieb, Matthias Lesch, Werner Müller, Joachim Schwermer, Gregor Weingart, and two anonymous referees for very helpful discussions and remarks about various topics of these notes. I would like to thank Anna Pratoussevitch, Daniel Roggenkamp, and Anna Wienhard for their careful proofreading of the manuscript. My special thanks go to Hans-Joachim Hein, who read many versions of the manuscript very carefully and suggested many substantial improvements. Subsections 4.6, 6.3, and 7.4 are taken from his Diplom thesis [Hei]. Finally I would like to thank Irene Zimmermann and Manfred Karbe from the EMS Publishing House for their cordial and effective cooperation.

Contents

1 Preliminaries

In this chapter, we set notation and conventions and discuss some preliminaries. Let M be a manifold[1] of dimension n. A *coordinate chart* for M is a tuple (x, U), where $U \subset M$ is open and $x\colon U \to \mathbb{R}^n$ is a diffeomorphism onto its image. As a rule, we will not refer to the domain U of x. The *coordinate frame* of a coordinate chart x consists of the ordered tuple of vector fields

$$X_j = \frac{\partial}{\partial x^j}, \qquad 1 \leq j \leq n. \tag{1.1}$$

We say that a coordinate chart x is *centered* at a point $p \in M$ if $x(p) = 0$.

Let $E \to M$ be a vector bundle over M. We denote the space of sections of E by $\mathcal{E}(E)$ or $\mathcal{E}(M, E)$. More generally, for $U \subset M$ open, we denote by $\mathcal{E}(U, E)$ the space of sections of E over U. Furthermore, we denote by $\mathcal{E}_c(U, E) \subset \mathcal{E}(U, E)$ the subspace of sections with compact support in U.

As long as there is no need to specify a name for them, Riemannian metrics on E are denoted by angle brackets $\langle \cdot, \cdot \rangle$. Similarly, if E is complex, Hermitian metrics on E are denoted by parentheses $(\cdot, \cdot)$. As a rule we assume that Hermitian metrics on a given bundle E are conjugate linear in the first variable and complex linear in the second. The induced Hermitian metric on the dual bundle E^* will then be complex linear in the first and conjugate linear in the second variable. The reason for using different symbols for Riemannian and Hermitian metrics is apparent from (1.12) below.

If E is a complex vector bundle over M and $(\cdot, \cdot)$ is a Hermitian metric on E, then

$$(\sigma, \tau)_2 = \int_M (\sigma, \tau) \tag{1.2}$$

is a Hermitian product on $\mathcal{E}_c(M, E)$. We let $L^2(E) = L^2(M, E)$ be the completion of $\mathcal{E}_c(M, E)$ with respect to the Hermitian norm induced by the Hermitian product in (1.2) and identify $L^2(M, E)$ with the space of equivalence classes of square-integrable measurable sections of E as usual[2].

Let $g = \langle \cdot, \cdot \rangle$ be a Riemannian metric on M and ∇ be its Levi-Civita connection. By setting $g(X)(Y) := g(X, Y)$, we may interpret g as an isomorphism $TM \to T^*M$. We use the standard musical notation for this isomorphism,

$$v^\flat(w) = \langle v, w \rangle \quad \text{and} \quad \langle \varphi^\sharp, w \rangle = \varphi(w), \tag{1.3}$$

where $v, w \in TM$ and $\varphi \in T^*M$ have the same foot points. It is obvious that $(v^\flat)^\sharp = v$ and $(\varphi^\sharp)^\flat = \varphi$.

For a vector field X, the Lie derivative $L_X g$ of g measures how much g varies under the flow of X. It is given by

$$(L_X g)(Y, Z) = \langle \nabla_Y X, Z \rangle + \langle Y, \nabla_Z X \rangle. \tag{1.4}$$

[1] Unless specified otherwise, manifolds and maps are assumed to be smooth.

[2] We use similar terminology in the case of real vector bundles and Riemannian metrics.

In fact, by the product rule for the Lie derivative,

$$X\langle Y, Z\rangle = L_X(g(Y, Z)) = (L_X g)(Y, Z) + g(L_X Y, Z) + g(Y, L_X Z).$$

We say that X is a *Killing field* if the flow of X preserves g. By definition, X is a Killing field iff $L_X g = 0$ or, equivalently, iff ∇X is skew-symmetric.

1.5 Exercises. Let X be a Killing field on M.

1) For any geodesic c in M, the composition $X \circ c$ is a Jacobi field along c. Hint: The flow of X consists of local isometries of M and gives rise to local geodesic variations of c with variation field $X \circ c$.

2) For all vector fields Y, Z on M

$$\nabla^2 X(Y, Z) + R(X, Y)Z = 0.$$

Hint: For $Y = Z$, this equation reduces to the Jacobi equation.

The *divergence* $\operatorname{div} X$ of a vector field X on M measures the change of the volume element under the flow of X. It is given by

$$\operatorname{div} X = \operatorname{tr} \nabla X. \tag{1.6}$$

In terms of a local orthonormal frame $(X_1, \dots, X_n)$ of TM, we have

$$\operatorname{div} X = \sum \langle \nabla_{X_j} X, X_j\rangle. \tag{1.7}$$

By (1.4), we also have

$$\operatorname{tr} L_X g = 2 \operatorname{div} X. \tag{1.8}$$

1.9 Divergence Formula. *Let G be a compact domain in M with smooth boundary ∂G and exterior normal vector field ν along ∂G. Then*

$$\int_G \operatorname{div} X = \int_{\partial G} \langle X, \nu\rangle.$$

1.1 Differential Forms. We let $A^*(M, \mathbb{R}) := \Lambda^*(T^*M)$ be the bundle of (multilinear) alternating forms on TM (with values in $\mathbb{R}$) and $\mathcal{A}^*(M, \mathbb{R}) := \mathcal{E}(A^*(M, \mathbb{R}))$ be the space of differential forms on M. We let $A^r(M, \mathbb{R})$ and $\mathcal{A}^r(M, \mathbb{R})$ be the subbundle of alternating forms of degree r and the subspace of differential forms on M of degree r, respectively.

The Riemannian metric on M induces a Riemannian metric on $A^*(M, \mathbb{R})$. Similarly, the Levi-Civita connection induces a connection $\hat{\nabla}$ on $A^*(M, \mathbb{R})$, compare the product rule 1.20 below, and $\hat{\nabla}$ is metric with respect to the induced metric on $A^*(M, \mathbb{R})$.

Recall the *interior product* of a tangent vector v with an alternating form φ with the same foot point, $v \llcorner \varphi = \varphi(v, \dots)$, that is, insert v as first variable. There are the following remarkable relations between $\wedge$ and $\llcorner$,

$$\langle v^\flat \wedge \varphi, \psi\rangle = \langle \varphi, v \llcorner \psi\rangle \tag{1.10}$$

and the *Clifford relation*

$$v^\flat \wedge (w \llcorner \varphi) + w \llcorner (v^\flat \wedge \varphi) = \langle v, w \rangle \varphi, \tag{1.11}$$

where v and w are tangent vectors and φ and ψ alternating forms, all with the same foot point.

We let $A^*(M, \mathbb{C}) := A^*(M, \mathbb{R}) \otimes_{\mathbb{R}} \mathbb{C}$ be the bundle of $\mathbb{R}$-multilinear alternating forms on TM with values in $\mathbb{C}$ and $\mathcal{A}^*(M, \mathbb{C}) := \mathcal{E}(A^*(M, \mathbb{C}))$ be the space of complex valued differential forms on M. Such forms decompose, $\varphi = \rho + i\tau$, where ρ and τ are differential forms with values in $\mathbb{R}$ as above. We call ρ the *real part*, $\rho = \operatorname{Re} \varphi$, and τ the *imaginary part*, $\tau = \operatorname{Im} \varphi$, of φ and set $\bar{\varphi} = \rho - i\tau$. Via complex multilinear extension, we may view elements of $A^*(M, \mathbb{C})$ as $\mathbb{C}$-multilinear alternating forms on $T_{\mathbb{C}} M = TM \otimes_{\mathbb{R}} \mathbb{C}$, the complexified tangent bundle.

We extend Riemannian metric, wedge product, and interior product complex linearly in the involved variables to $T_{\mathbb{C}} M$ and $A^*(M, \mathbb{C})$, respectively. We extend $\hat{\nabla}$ complex linearly to a connection on $A^*(M, \mathbb{C})$, $\hat{\nabla}\varphi := \hat{\nabla} \operatorname{Re} \varphi + i \hat{\nabla} \operatorname{Im} \varphi$. Equations 1.10 and 1.11 continue to hold, but now with complex tangent vectors v, w and $\mathbb{C}$-valued forms φ and ψ. There is an induced Hermitian metric

$$(\varphi, \psi) := \langle \bar{\varphi}, \psi \rangle \tag{1.12}$$

on $A^*(M, \mathbb{C})$ and a corresponding L^2-Hermitian product on $\mathcal{A}_c^*(M, \mathbb{C})$ as in (1.2).

Let $E \to M$ be a complex vector bundle. A differential form with values in E is a smooth section of $A^*(M, E) := A^*(M, \mathbb{C}) \otimes E$, that is, an element of $\mathcal{A}^*(M, E) := \mathcal{E}(A^*(M, E))$. Locally, any such form is a linear combination of *decomposable* differential forms $\varphi \otimes \sigma$ with $\varphi \in \mathcal{A}^*(M, \mathbb{C})$ and $\sigma \in \mathcal{E}(M, E)$. We define the wedge product of $\varphi \in \mathcal{A}^*(M, \mathbb{C})$ with $\alpha \in \mathcal{A}^*(M, E)$ by

$$\varphi \wedge \alpha := \sum (\varphi \wedge \varphi_j) \otimes \sigma_j, \tag{1.13}$$

where we decompose $\alpha = \sum \varphi_j \otimes \sigma_j$. More generally, let E' and E'' be further complex vector bundles over M and $\mu \colon E \otimes E' \to E''$ be a morphism. We define the wedge product of differential forms $\alpha \in \mathcal{A}^*(M, E)$ and $\alpha' \in \mathcal{A}^*(M, E')$ by

$$\alpha \wedge_\mu \alpha' := \sum_{j,k} (\varphi_j \wedge \varphi_k') \otimes \mu(\sigma_j \otimes \sigma_k') \in \mathcal{A}(M, E''), \tag{1.14}$$

where we write $\alpha = \sum \varphi_j \otimes \sigma_j$ and $\alpha' = \sum \varphi_k' \otimes \sigma_k'$. If α and α' are of degree r and s, respectively, then

$$\begin{aligned}
(\alpha \wedge_\mu \alpha')&(X_1, \dots, X_{r+s}) \\
&= \frac{1}{r!s!} \sum \varepsilon(\sigma) \mu\{\alpha(X_{\sigma(1)}, \dots, X_{\sigma(r)}) \otimes \alpha'(X_{\sigma(r+1)}, \dots, X_{\sigma(r+s)})\}.
\end{aligned} \tag{1.15}$$

This formula shows that the wedge products in (1.13) and (1.14) do not depend on the way in which we write α and α' as sums of decomposable differential forms.

1.16 Exercise. Let E be a complex vector bundle over M and μ and λ be composition and Lie bracket in the associated vector bundle End E of endomorphisms of E. Let α, α', and α'' be differential forms with values in End E. Then

$$(\alpha \wedge_\mu \alpha') \wedge_\mu \alpha'' = \alpha \wedge_\mu (\alpha' \wedge_\mu \alpha'')$$

and

$$\alpha \wedge_\lambda \alpha' = \alpha \wedge_\mu \alpha' - (-1)^{rs} \alpha' \wedge_\mu \alpha,$$

if α and α' have degree r and s, respectively.

1.2 Exterior Derivative. We refer to the beginning of Appendix C for some of the terminology in this and the next subsection. Let E be a complex vector bundle over M and D be a connection on E. For a differential form α with values in E and of degree r, we define the exterior derivative of α (with respect to D) by

$$d^D\alpha := \sum (d\varphi_j \otimes \sigma_j + (-1)^r \varphi_j \wedge D\sigma_j), \tag{1.17}$$

where we decompose $\alpha = \sum \varphi_j \otimes \sigma_j$ and where d denotes the usual exterior derivative. That this is independent of the decomposition of α into a sum of decomposable differential forms follows from

$$\begin{aligned} d^D\alpha(X_0, \dots, X_r) = &\sum_j (-1)^j D_{X_j}(\alpha(X_0, \dots, \hat{X}_j, \dots, X_r)) \\ &+ \sum_{j<k} (-1)^{j+k} \alpha([X_j, X_k], X_0, \dots, \hat{X}_j, \dots, \hat{X}_k, \dots, X_r). \end{aligned} \tag{1.18}$$

Since $d^D(f\alpha) - f d^D\alpha = df \wedge \alpha$ for any function f on M, the principal symbol $\sigma\colon T^*M \otimes A^*(M, E) \to A^*(M, E)$ of d^D is given by $\sigma(\xi \otimes \alpha) = \xi \wedge \alpha$.

1.19 Proposition. *Let E be a vector bundle over M and D be a connection on E. Then, for all $\alpha \in \mathcal{A}^*(M, E)$,*

$$d^D d^D \alpha = R^D \wedge_\varepsilon \alpha,$$

where we view the curvature tensor R^D of D as a two-form with values in the bundle End E *of endomorphisms of E and where the wedge product is taken with respect to the evaluation map ε:* End $E \otimes E \to E$. □

Via the product rule, $A^*(M, E)$ inherits a connection $\hat{D}$ from $\hat{\nabla}$ and D,

$$\hat{D}_X(\varphi \otimes \sigma) := (\hat{\nabla}_X \varphi) \otimes \sigma + \varphi \otimes (D_X \sigma). \tag{1.20}$$

In terms of $\hat{D}$, we have

$$d^D\alpha = \sum X_j^* \wedge \hat{D}_{X_j}\alpha, \tag{1.21}$$

where $(X_1, \dots, X_n)$ is a local frame of TM and $(X_1^*, \dots, X_n^*)$ is the corresponding dual frame. We emphasize that the curvature tensors of $\hat{\nabla}$ and $\hat{D}$ act as derivations with respect to wedge and tensor products.

Let E, E', E'' and $\mu\colon E \otimes E' \to E''$ be as in (1.14). Let D, D' and D'' be connections on E, E', and E'', respectively, such that the product rule

$$D''_X(\mu(\sigma \otimes \tau)) = \mu((D'_X\sigma) \otimes \tau) + \mu(\sigma \otimes (D''_X\tau)) \tag{1.22}$$

holds for all vector fields X of M. The induced connections on the bundles of forms then also satisfy the corresponding product rule,

$$\hat{D}''_X(\alpha \wedge_\mu \alpha') = (\hat{D}_X\alpha) \wedge_\mu \alpha' + (\alpha \wedge_\mu (\hat{D}'_X\alpha')). \tag{1.23}$$

For the exterior differential we get

$$d^{D''}(\alpha \wedge_\mu \alpha') = (d^D\alpha) \wedge_\mu \alpha' + (-1)^r\alpha \wedge_\mu (d^{D'}\alpha'), \tag{1.24}$$

where we assume that α is of degree r.

Let $h = (\cdot, \cdot)$ be a Hermitian metric on E. Then h induces a Hermitian metric on $A^*(M, E)$, on decomposable forms given by

$$(\varphi \otimes \sigma, \psi \otimes \tau) = \langle \bar{\varphi}, \psi \rangle (\sigma, \tau), \tag{1.25}$$

and a corresponding L^2-Hermitian product on $\mathcal{A}_c^*(M, E)$ as in (1.2).

1.26 Exercise. The analogs of (1.10) and (1.11) hold on $A^*(M, E)$,

$$(v^\flat \wedge \alpha, \beta) = (\alpha, v\llcorner\beta) \quad \text{and} \quad v^\flat \wedge (w\llcorner\alpha) + w\llcorner(v^\flat \wedge \alpha) = \langle v, w\rangle\alpha,$$

where $v, w \in TM$ and $\alpha, \beta \in A^*(M, E)$ have the same foot point.

Let D be a Hermitian connection on E. Then the induced connection $\hat{D}$ on $A^*(M, E)$ as in (1.20) is Hermitian as well.

1.27 Proposition. *In terms of a local orthonormal frame $(X_1, \dots, X_n)$ of M, the formal adjoint $(d^D)^*$ of d^D is given by*

$$(d^D)^*\alpha = -\sum X_j\llcorner\hat{D}_{X_j}\alpha.$$

Proof. Let α and β be differential forms of degree $r-1$ and r. Let $p \in M$ and choose a local orthonormal frame (X_j) around p with $\nabla X_j(p) = 0$. Then, at p,

$$\begin{aligned}(d^D\alpha, \beta) &= \sum (X_j^* \wedge \hat{D}_{X_j}\alpha, \beta)\\ &= \sum (\hat{D}_{X_j}\alpha, X_j\llcorner\beta)\\ &= \sum X_j(\alpha, X_j\llcorner\beta) - \sum(\alpha, X_j\llcorner\hat{D}_{X_j}\beta).\end{aligned}$$

The first term on the right is equal to the divergence of the complex vector field Z defined by $(Z, W) = (\alpha, W\llcorner\beta)$, see (C.3). □

Since $(d^D)^*(f\alpha) - f(d^D)^*\alpha = -\operatorname{grad} f\llcorner\alpha$ for any function f on M, the principal symbol $\sigma\colon T^*M \otimes A^*(M, E) \to A^*(M, E)$ of $(d^D)^*$ is given by $\sigma(\xi \otimes \alpha) = -\xi^\sharp\llcorner\alpha$.

1.3 Laplace Operator. As above, we let $E \to M$ be a complex vector bundle with Hermitian metric h and Hermitian connection D. We say that a differential form α with values in E is *harmonic* if $d^D\alpha = (d^D)^*\alpha = 0$ and denote by $\mathcal{H}^*(M, E)$ the space of harmonic differential forms with values in E.

The *Laplace operator* associated to d^D is

$$\Delta_{d^D} = d^D(d^D)^* + (d^D)^*d^D. \tag{1.28}$$

Since the principal symbol of a composition of differential operators is the composition of their principal symbols, the principal symbol σ of Δ_{d^D} is given by

$$\sigma(\xi \otimes \alpha) = -(\xi \wedge (\xi^\sharp \llcorner \alpha) + \xi^\sharp \llcorner (\xi \wedge \alpha)) = -||\xi||^2\alpha, \tag{1.29}$$

by (1.11). In particular, Δ_{d^D} is an elliptic differential operator.

1.30 Exercise. Assume that M is closed. Use the divergence formula (1.9) to show that

$$(\Delta_{d^D}\alpha, \beta)_2 = (d^D\alpha, d^D\beta)_2 + ((d^D)^*\alpha, (d^D)^*\beta)_2.$$

Conclude that φ is harmonic iff $\Delta_{d^D}\alpha = 0$. Compare also Corollary C.22.

Using the Clifford relation 1.11 and the formulas 1.21 and 1.27 for d^D and $(d^D)^*$, a straightforward calculation gives the *Weitzenböck formula*

$$\Delta_{d^D}\alpha = -\sum_j \hat{D}^2\alpha(X_j, X_j) + \sum_{j \neq k} X_k^* \wedge (X_j \llcorner \hat{R}^D(X_j, X_k)\alpha), \tag{1.31}$$

where $(X_1, \ldots, X_n)$ is a local orthonormal frame of TM, $(X_1^*, \ldots, X_n^*)$ is the corresponding dual frame of T^*M, and $\hat{R}^D$ denotes the curvature tensor of the connection $\hat{D}$ on $A^*(M, E)$. Denoting the second term on the right hand side of (1.31) by $K\alpha$ and using (C.8), we can rewrite (1.31) in two ways,

$$\Delta_{d^D}\alpha = -\operatorname{tr}\hat{D}^2\alpha + K\alpha = \hat{D}^*\hat{D}\alpha + K\alpha. \tag{1.32}$$

Assume now that D is flat. If (Φ_j) is a parallel frame of E over an open subset U of M, then over U, $\alpha \in \mathcal{A}(M, E)$ can be written as a sum $\alpha = \sum \varphi_j \otimes \Phi_j$ and

$$d^D\alpha = \sum (d\varphi_j) \otimes \Phi_j \quad \text{and} \quad (d^D)^*\alpha = \sum (d^*\varphi_j) \otimes \Phi_j. \tag{1.33}$$

For that reason, we often use the shorthand $d\alpha$ and $d^*\alpha$ for $d^D\alpha$ and $(d^D)^*\alpha$ if D is flat. Then we also have $d^2 = (d^*)^2 = 0$, see Proposition 1.19. The latter implies that the images of d and d^* are L^2-perpendicular and that the Laplace operator is a square,

$$\Delta_d = (d + d^*)^2. \tag{1.34}$$

The fundamental estimates and regularity theory for elliptic differential operators lead to the Hodge decomposition of $\mathcal{A}^*(M, E)$, see for example Section III.5 in [LM] or Chapter 6 in [Wa]:

1.35 Theorem (Hodge Decomposition). *If M is closed and D is flat, then*

$$\mathcal{A}^*(M, E) = \mathcal{H}^*(M, E) + d(\mathcal{A}^*(M, E)) + d^*(\mathcal{A}^*(M, E)),$$

where the sum is orthogonal with respect to the L^2-Hermitian product on $\mathcal{A}^*(M, E)$.

In particular, if M is closed and D is flat, then the canonical map to cohomology,

$$\mathcal{H}^*(M, E) \to H^*(M, E), \tag{1.36}$$

is an isomorphism of vector spaces. In other words, each de Rham cohomology class of M with coefficients in E contains precisely one harmonic representative. In particular, $\dim H^*(M, E) < \infty$. The most important case is $E = \mathbb{C}$.

1.37 Remark. It is somewhat tempting to assume that the ring structure of $H^*(M, \mathbb{C})$ is also represented by $\mathcal{H}^*(M, \mathbb{C})$. However, this only happens in rare cases. As a rule, the wedge product of harmonic differential forms is not a harmonic differential form anymore. In fact, it is a specific property of the Kähler form of a Kähler manifold that its wedge product with a harmonic form gives a harmonic form, see Theorem 5.25. Compare also Remark 1.42 below.

In the above discussion, we only considered complex vector bundles. There is a corresponding theory in the real case, which we will use in some instances.

1.38 Exercise. Show that for $\varphi \in \mathcal{A}^1(M, \mathbb{C})$, the curvature term K in the Weitzenböck formula (1.32) is given by $K\varphi = (\operatorname{Ric}\varphi^\sharp)^\flat$, where Ric denotes the Ricci tensor of M. In other words, $\Delta_d\varphi = \hat{\nabla}^*\hat{\nabla}\varphi + (\operatorname{Ric}\varphi^\sharp)^\flat$.

The equation in Exercise 1.38 was observed by Bochner (see also [Ya]). In the following theorem we give his ingenious application of it. Bochner's argument can be used in many other situations and is therefore named after him. Let $b_j(M)$ be the j-th Betti number of M, $b_j(M) = \dim_{\mathbb{R}} H^j(M, \mathbb{R}) = \dim_{\mathbb{C}} H^j(M, \mathbb{C})$.

1.39 Theorem (Bochner [Boc]). *Let M be a closed and connected Riemannian manifold with non-negative Ricci curvature. Then $b_1(M) \le n$ with equality if and only if M is a flat torus. If, in addition, the Ricci curvature of M is positive in some point of M, then $b_1(M) = 0$.*

Proof. Since M is closed, we can represent real cohomology classes of dimension one uniquely by harmonic one-forms, by the (identical) version of Theorem 1.35 for real vector bundles. If φ is such a differential form, then

$$0 = \int_M \langle \Delta_d\varphi, \varphi\rangle = \int_M ||\hat{\nabla}\varphi||^2 + \int \langle \operatorname{Ric}\varphi^\sharp, \varphi^\sharp\rangle,$$

by Exercise 1.38. By assumption, the integrand in the second integral on the right is non-negative. It follows that φ is parallel, therefore also the vector field $\varphi^\sharp$, and that $\operatorname{Ric}\varphi^\sharp = 0$. The rest of the argument is left as an exercise. □

1.40 Remarks. 1) The earlier theorem of Bonnet–Myers makes the stronger assertion that the fundamental group of a closed, connected Riemannian manifold with positive Ricci curvature is finite.

2) Let M be a closed and connected Riemannian manifold with non-negative Ricci curvature. If the Ricci curvature of M is positive in some point of M, then the Riemannian metric of M can be deformed to a Riemannian metric of positive Ricci curvature [Au1] (see also [Eh]).

3) The complete analysis of the fundamental groups of closed and connected Riemannian manifolds with non-negative Ricci curvature was achieved by Cheeger and Gromoll [CG1], [CG2]. Compare Subsection 6.1.

1.41 Exercise. Conclude from the argument in the proof of Theorem 1.39 that a closed, connected Riemannian manifold M with non-negative Ricci curvature is foliated by a parallel family of totally geodesic flat tori of dimension $b_1(M)$. Hint: The space $\mathfrak{p}$ of parallel vector fields on M is an Abelian subalgebra of the Lie algebra of Killing fields on M. The corresponding connected subgroup of the isometry group of M is closed and Abelian and its orbits foliate M by parallel flat tori.

1.42 Remark (and Exercise). The curvature operator $\hat{R}$ is the symmetric endomorphism on $\Lambda^2(TM)$ defined by the equation

$$\langle \hat{R}(X \wedge Y), U \wedge V\rangle := \langle R(X,Y)V, U\rangle. \tag{1.43}$$

Gallot and Meyer showed that the curvature term in the Weitzenböck formula (1.32) for Δ_d on $\mathcal{A}^*(M,\mathbb{R})$ is positive or non-negative if $\hat{R} > 0$ or $\hat{R} \geq 0$, respectively, see [GM] or (the proof of) Theorem 8.6 in [LM]. In particular, if M is closed with $\hat{R} > 0$, then $b_r(M) = 0$ for $0 < r < n$, by Hodge theory as in (1.36) and the Bochner argument in Theorem 1.39. If M is closed with $\hat{R} \geq 0$, then real valued harmonic forms on M are parallel, again by the Bochner argument. Since the wedge product of parallel differential forms is a parallel differential form, hence a harmonic form, this is one of the rare instances where the wedge product of harmonic forms is harmonic (albeit for a trivial reason), compare Remark 1.37. If M is also connected, then a parallel differential form on M is determined by its value at any given point of M and hence $b_r(M) \leq \binom{n}{r}$ for $0 < r < n$. Equality for any such r implies that M is a flat torus.

1.4 Hodge Operator. Suppose now that M is oriented, and denote by vol the oriented volume form of M. Then the Hodge operator $*$ is defined[3] by the tensorial equation

$$*\varphi \wedge \psi = \langle \varphi, \psi\rangle \,\mathrm{vol}. \tag{1.44}$$

By definition, we have

$$*\,1 = \mathrm{vol}, \quad *\,\mathrm{vol} = 1 \quad \text{and} \quad **= \sum (-1)^{r(n-r)} P_r, \tag{1.45}$$

[3]Note that the definition here differs from the standard one, the definition here gives a more convenient sign in (1.47).

where $P_r\colon \mathcal{A}^*(M,\mathbb{R}) \to \mathcal{A}^r(M,\mathbb{R})$ is the natural projection.

Let φ and ψ be differential forms with compact support and of degree r and $r-1$, respectively. Since $d*\varphi$ is a differential form of degree $n-r+1$ and $r-r^2$ is even, we have

$$**d*\varphi = (-1)^{(n-r+1)(r-1)} d*\varphi = (-1)^{nr+n-r+1} d*\varphi.$$

Hence

$$\begin{aligned} d(*\varphi \wedge \psi) &= d*\varphi \wedge \psi + (-1)^{n-r} *\varphi \wedge d\psi \\ &= (-1)^{nr+n-r+1} *(*d*\varphi) \wedge \psi + (-1)^{n-r} *\varphi \wedge d\psi \\ &= (-1)^{n-r}\{(-1)^{nr+1}\langle *d*\varphi, \psi\rangle + \langle \varphi, d\psi\rangle\} \operatorname{vol}. \end{aligned} \tag{1.46}$$

By Stokes' theorem, the integral over M of the left hand side vanishes. We conclude that on differential forms of degree r

$$d^* = (-1)^{nr} *d* . \tag{1.47}$$

It is now easy to check that

$$*\Delta_d = \Delta_d * . \tag{1.48}$$

If M is closed, then $*$ maps harmonic forms to harmonic forms, by Exercise 1.30 and Equation 1.48. Hence, for closed M,

$$*\colon \mathcal{H}^r(M,\mathbb{R}) \to \mathcal{H}^{n-r}(M,\mathbb{R}) \tag{1.49}$$

is an isomorphism. This is Poincaré duality on the level of harmonic forms.

Extend $*$ complex linearly to $A^*(M,\mathbb{C})$. Let E be a vector bundle over M and E^* be the dual bundle of E. Assume that E is endowed with a Hermitian metric. Via $h(\sigma)(\tau) = (\sigma,\tau)$ view the Hermitian metric of E as a conjugate linear isomorphism $h\colon E \to E^*$. We obtain a conjugate linear isomorphism

$$\bar{*} \otimes h\colon \mathcal{A}^*(M,E) \to \mathcal{A}^*(M,E^*), \tag{1.50}$$

where $\bar{*}\varphi := \overline{*\varphi}$. We have

$$((\bar{*}\otimes h)\alpha) \wedge_\varepsilon \beta = (\alpha,\beta) \operatorname{vol}, \tag{1.51}$$

where $\varepsilon\colon E^* \otimes E \to \mathbb{C}$ is the evaluation map. Note that $((\bar{*}\otimes h)\alpha) \wedge_\varepsilon \beta$ is complex valued.

Let D^* be the induced connection on E^*. With respect to D and D^*, h is a parallel morphism from E to E^*,

$$(D^*_X(h(\sigma)))(\tau) = X(\sigma,\tau) - (\sigma, D_X\tau) = (D_X\sigma,\tau) = h(D_X\sigma)(\tau).$$

The induced (conjugate) Hermitian metric h^* on E^* is given by

$$(h(\sigma), h(\tau)) := (\sigma,\tau). \tag{1.52}$$

Note that h^* is complex linear in the first and conjugate linear in the second variable. The connection D^* is Hermitian with respect to h^*,

$$\begin{aligned} X(h(\sigma), h(\tau)) &= X(\sigma, \tau) = (D_X\sigma, \tau) + (\sigma, D_X\tau) \\ &= (h(D_X\sigma), h(\tau)) + (h(\sigma), h(D_X\tau)) \\ &= (D^*(h(\sigma)), h(\tau)) + (h(\sigma), D^*_X(h(\tau)). \end{aligned}$$

Via $\xi(h^*(\eta)) = (\xi, \eta)$ we consider h^* as a conjugate linear isomorphism from E^* to E. We have

$$(\sigma, h^*(h(\tau))) = h(\sigma)(h^*(h(\tau))) = (h(\sigma), h(\tau)) = (\sigma, \tau),$$

and hence $h^*h = \mathrm{id}$. It follows that

$$(\bar{*} \otimes h^*)(\bar{*} \otimes h) = (\bar{*} \otimes h)(\bar{*} \otimes h^*) = (-1)^{r(n-r)} \tag{1.53}$$

on forms of degree r and $n-r$. Using (1.24) with D'' the usual derivative of functions and computing as in (1.46), we get

$$(d^D)^* = (-1)^{nr}(\bar{*} \otimes h^*)\, d^{D^*} (\bar{*} \otimes h). \tag{1.54}$$

This implies that the corresponding Laplacians, for simplicity denoted Δ, satisfy

$$\Delta\,(\bar{*} \otimes h) = (\bar{*} \otimes h)\,\Delta. \tag{1.55}$$

If M is closed, then $\bar{*} \otimes h$ maps harmonic forms to harmonic forms, by Exercise 1.30 and Equation 1.55. Hence, for closed M, $\bar{*} \otimes h$ induces conjugate linear isomorphisms

$$\mathcal{H}^r(M, E) \to \mathcal{H}^{n-r}(M, E^*). \tag{1.56}$$

This is Poincaré duality for vector bundle valued harmonic forms.

2 Complex Manifolds

Let V be a vector space over $\mathbb{R}$. A *complex structure* on V is an endomorphism $J\colon V \to V$ such that $J^2 = -1$. Such a structure turns V into a complex vector space by defining multiplication with i by $iv := Jv$. Vice versa, multiplication by i in a complex vector space is a complex structure on the underlying real vector space.

2.1 Example. To fix one of our conventions, we discuss the complex vector space $\mathbb{C}^m$ explicitly. Write a vector in $\mathbb{C}^m$ as a tuple

$$(z^1, \dots, z^m) = (x^1 + iy^1, \dots, x^m + iy^m)$$

and identify it with the vector $(x^1, y^1, \dots, x^m, y^m)$ in $\mathbb{R}^{2m}$. The corresponding complex structure on $\mathbb{R}^{2m}$ is

$$J(x^1, y^1, \dots, x^m, y^m) = (-y^1, x^1, \dots, -y^m, x^m).$$

We will use this identification of $\mathbb{C}^m$ with $\mathbb{R}^{2m}$ and complex structure J on $\mathbb{R}^{2m}$ without further reference.

Let M be a smooth manifold of real dimension $2m$. We say that a smooth atlas $\mathcal{A}$ of M is *holomorphic* if for any two coordinate charts $z\colon U \to U' \subset \mathbb{C}^m$ and $w\colon V \to V' \subset \mathbb{C}^m$ in $\mathcal{A}$, the coordinate transition map $z \circ w^{-1}$ is holomorphic. Any holomorphic atlas uniquely determines a maximal holomorphic atlas, and a maximal holomorphic atlas is called a *complex structure*. We say that M is a *complex manifold* of complex dimension m if M comes equipped with a holomorphic atlas. Any coordinate chart of the corresponding complex structure will be called a *holomorphic coordinate chart* of M. A *Riemann surface* or *complex curve* is a complex manifold of complex dimension 1.

Let M be a complex manifold. Then the transition maps $z \circ w^{-1}$ of holomorphic coordinate charts are biholomorphic. Hence they are diffeomorphisms and the determinants of their derivatives, viewed as $\mathbb{R}$-linear maps, are positive. It follows that a holomorphic structure determines an orientation of M, where we choose $dx^1 \wedge dy^1 \wedge \dots \wedge dx^m \wedge dy^m$ as orientation of $\mathbb{C}^m$, compare Example 2.1.

We say that a map $f\colon M \to N$ between complex manifolds is *holomorphic* if, for all holomorphic coordinate charts $z\colon U \to U'$ of M and $w\colon V \to V'$ of N, the map $w \circ f \circ z^{-1}$ is holomorphic on its domain of definition. We say that f is *biholomorphic* if f is bijective and f and f^{-1} are holomorphic. An *automorphism* of a complex manifold M is a biholomorphic map $f\colon M \to M$.

To be consistent in what we say next, we remark that open subsets of complex manifolds inherit a complex structure. For an open subset U of a complex manifold M, we denote by $\mathcal{O}(U)$ the set of holomorphic functions $f\colon U \to \mathbb{C}$, a ring under pointwise addition and multiplication of functions.

The inverse mapping and implicit function theorem also hold in the holomorphic setting. Corresponding to the real case, we have the notions of holomorphic immersion, holomorphic embedding, and complex submanifold. The discussion is completely parallel to the discussion in the real case.

Of course, complex analysis is different from real analysis. To state just one phenomenon where they differ, by the maximum principle a holomorphic function on a closed complex manifold is locally constant. In particular, $\mathbb{C}^m$ does not contain closed complex submanifolds (of positive dimension).

2.2 Examples. 1) Let $U \subset \mathbb{C}^m$ be an open subset. Then M together with the atlas consisting of the one coordinate chart $\mathrm{id}\colon U \to U$ is a complex manifold.

2) *Riemann sphere.* Consider the unit sphere

$$S^2 = \{(w, h) \in \mathbb{C} \times \mathbb{R} \mid w\bar{w} + h^2 = 1\}.$$

Let $N = (0, 1)$ and $S = (0, -1)$ be the north and south pole of S^2, respectively. The stereographic projections $\pi_N\colon S^2 \setminus \{N\} \to \mathbb{C}$ and $\pi_S\colon S^2 \setminus \{S\} \to \mathbb{C}$ are given by $\pi_N(w, h) = (1 - h)^{-1}w$ and $\pi_S(w, h) = (1 + h)^{-1}w$, respectively. The transition map $\pi_S \circ \pi_N^{-1}\colon \mathbb{C} \setminus \{0\} \to \mathbb{C} \setminus \{0\}$ is given by $(\pi_S \circ \pi_N^{-1})(z) = 1/\bar{z}$. It is smooth, and thus π_N and π_S define a smooth atlas of S^2. However, it is not holomorphic. We obtain a holomorphic atlas by replacing π_S by its complex conjugate, $\bar{\pi}_S$. Then the transition map is $(\bar{\pi}_S \circ \pi_N^{-1})(z) = 1/z$, and hence the atlas of S^2 consisting of π_N and $\bar{\pi}_S$ is holomorphic. The Riemann sphere is S^2 together with the complex structure determined by this atlas. It is a consequence of the uniformization theorem that this complex structure on S^2 is unique up to diffeomorphism. As we will see, the Riemann sphere is biholomorphic to the complex line $\mathbb{C}P^1$, described in the next example.

3) *Complex projective spaces.* As a set, complex projective space $\mathbb{C}P^m$ is the space of all complex lines in $\mathbb{C}^{m+1}$. For a non-zero vector $z = (z^0, \dots, z^m) \in \mathbb{C}^{m+1}$, we denote by $[z]$ the complex line generated by z and call $(z^0, \dots, z^m)$ the *homogeneous coordinates* of $[z]$. For $0 \le j \le m$, we let $U_j = \{[z] \in \mathbb{C}P^m \mid z^j \neq 0\}$. Each $[z]$ in U_j intersects the affine hyperplane $\{z^j = 1\}$ in $\mathbb{C}^{m+1}$ in exactly one point. We use this to obtain a coordinate map

$$a_j\colon U_j \to \mathbb{C}^m, \quad a_j([z]) = \frac{1}{z^j}(z^0, \dots, \hat{z}^j, \dots, z^m),$$

where the hat indicates that z^j is to be deleted. By what we said it is clear that a_j is a bijection. For $j < k$, the transition map $a_j \circ a_k^{-1}$ is defined on $\{w \in \mathbb{C}^m \mid w^j \neq 0\}$ and given by inserting 1 as k-th variable, multiplying the resulting $(m + 1)$-vector by $(w^j)^{-1}$, and deleting the redundant j-th variable 1. Thus the transition maps are holomorphic. It is now an exercise to show that there is precisely one topology on $\mathbb{C}P^m$ such that the maps a_j are coordinate charts and such that $\mathbb{C}P^m$ together with this topology and the atlas of maps a_j is a complex manifold of complex dimension m. For $m = 1$, we speak of the *complex projective line*, for $m = 2$ of the *complex projective plane*.

For $m \le n$, the map $f\colon \mathbb{C}P^m \to \mathbb{C}P^n$, $[z] \mapsto [z, 0]$, is a holomorphic embedding. More generally, if $A\colon \mathbb{C}^{m+1} \to \mathbb{C}^{n+1}$ is an injective linear map, then the induced map

$f: \mathbb{C}P^m \to \mathbb{C}P^n$, $[z] \mapsto [Az]$, is a holomorphic embedding. Thus we can view $\mathbb{C}P^m$ in many different ways as a complex submanifold of $\mathbb{C}P^n$. We also conclude that the group $\mathrm{PGl}(m+1, \mathbb{C}) = \mathrm{Gl}(m+1, \mathbb{C})/\mathbb{C}^*$ acts by biholomorphic transformations on $\mathbb{C}P^m$. It is known that $\mathrm{PGl}(m+1, \mathbb{C})$ is actually equal to the group of biholomorphic transformations of $\mathbb{C}P^m$.

For the Riemann sphere S^2 as in the previous example, the map $f: S^2 \to \mathbb{C}P^1$,

$$f(p) = \begin{cases} [\pi_N(p), 1] & \text{if } p \neq N, \\ [1, \bar{\pi}_S(p)] & \text{if } p \neq S, \end{cases}$$

is well defined and biholomorphic and thus identifies the Riemann sphere with the complex projective line.

4) *Complex Grassmannians.* This example generalizes the previous one. Let V be a complex vector space of dimension n and $G_r V$ be the space of r-dimensional complex subspaces of V, where $0 < r < n$.

Let M^* be the set of linear maps $F: \mathbb{C}^r \to V$ of rank r. There is a canonical projection

$$\pi: M^* \to G_r V, \quad \pi(F) = [F] =: \operatorname{im} F.$$

Let $B: V \to \mathbb{C}^n$ be an isomorphism. Then the map

$$M^* \to \mathbb{C}^{n \times r}, \quad F \mapsto \mathrm{Mat}(BF),$$

where $\mathrm{Mat}(BF)$ denotes the matrix of the linear map $BF: \mathbb{C}^r \to \mathbb{C}^n$, is a bijection onto the open subset of $(n \times r)$-matrices of rank r. This turns M^* into a complex manifold of dimension nr, and the complex structure on M^* does not depend on the choice of B. We write

$$\mathrm{Mat}(BF) = \begin{pmatrix} F_0 \\ F_1 \end{pmatrix}, \quad \text{where } F_0 \in \mathbb{C}^{r \times r} \text{ and } F_1 \in \mathbb{C}^{(n-r) \times r},$$

and let U_B be the subset of $[F]$ in $G_r V$ such that F_0 has rank r. We leave it as an exercise to the reader to show that

$$Z_B: U_B \to \mathbb{C}^{(n-r) \times r}, \quad Z_B([F]) = F_1 F_0^{-1},$$

is a well defined bijection and that, for any two isomorphisms $B, C: V \to \mathbb{C}^n$, the transition map $Z_B \circ Z_C^{-1}$ is holomorphic. With the same arguments as in the previous example we get that $G_r V$ has a unique topology such that the maps Z_B are coordinate charts turning $G_r V$ into a complex manifold of complex dimension $r(n-r)$. Moreover, $\pi: M^* \to G_r V$ is a holomorphic submersion and, for any isomorphism $B: V \to \mathbb{C}^n$,

$$\varphi_B: U_B \to M^*, \quad \varphi_B([F]) = B^{-1} \begin{pmatrix} 1 \\ F_1 F_0^{-1} \end{pmatrix},$$

where 1 stands for the $r \times r$ unit matrix and where we consider the matrix on the right as a linear map $\mathbb{C}^r \to \mathbb{C}^n$, is a holomorphic section of π.

The group $\mathrm{Gl}(r, \mathbb{C})$ of invertible matrices in $\mathbb{C}^{r\times r}$ is a complex Lie group (for complex Lie groups, see Example 8 below) and, considered as group of automorphisms of $\mathbb{C}^r$, acts on M^* on the right,

$$M^* \times \mathrm{Gl}(r, \mathbb{C}) \to M^*, \quad (F, A) \mapsto FA.$$

This action is holomorphic and turns $\pi\colon M^* \to G_r V$ into a principal bundle with structure group $\mathrm{Gl}(r, \mathbb{C})$. The complex Lie group $\mathrm{Gl}(V)$ of automorphisms of V acts on M^* and $G_r V$ on the left,

$$\mathrm{Gl}(V) \times M^* \to M^*, \quad (A, F) \mapsto AF,$$

respectively

$$\mathrm{Gl}(V) \times G_r V \to G_r V, \quad (A, [F]) \mapsto [AF].$$

These actions are also holomorphic and π is equivariant with respect to them.

5) *Tautological or universal bundle.* Let $0 < r < n$ and $M = G_r V$ be the Grassmannian of r-dimensional complex linear subspaces in V as in the previous example. As a set, the universal bundle over $G_r V$ is equal to

$$U_r V = \{(W, w) \mid W \in G_r V, w \in W\}.$$

There is a canonical projection

$$\pi\colon U_r V \to G_r V, \quad (W, w) \mapsto W.$$

For each $W \in G_r V$, the bijection $\pi^{-1}(W) \ni (W, w) \mapsto w \in W$ turns the fiber $\pi^{-1}(W)$ into an r-dimensional complex vector space isomorphic to W.

Let $B\colon V \to \mathbb{C}^n$ be an isomorphism and $U_B \subset G_r V$ be as in the previous example. Define a bijection

$$\Phi_B\colon U_B \times \mathbb{C}^r \to \pi^{-1}(U_B), \quad \Phi_B([F], v) = ([F], \varphi_B([F])v),$$

where φ_B is as in the previous example. For each $[F] \in U_B$, the map

$$\mathbb{C}^r \to \pi^{-1}([F]), \quad v \mapsto \Phi_B([F], v),$$

is an isomorphism of vector spaces. With arguments similar to the ones in the previous examples it follows that $U_r V$ is a complex manifold in a unique way such that $\pi\colon U_r V \to G_r V$ is a complex vector bundle over $G_r V$ and such that the trivializations Φ_B as above are holomorphic. In particular, the complex dimension of $U_r V$ is $r(n-r+1)$ and π is holomorphic. Moreover, the left action of $\mathrm{Gl}(V)$ on $G_r V$ extends canonically to a holomorphic action on $U_r V$,

$$\mathrm{Gl}(V) \times U_r V \to U_r V, \quad (A, (W, w)) \mapsto (AW, Aw).$$

This action has two orbits: the set of pairs (W, w) with $w \neq 0$ and of pairs $(W, 0)$.

6) *Complex tori.* Choose an $\mathbb{R}$-basis $B = (b_1, \dots, b_{2m})$ of $\mathbb{C}^m$. Let $\Gamma \subset \mathbb{C}^m$ be the *lattice* consisting of all integral linear combinations of B, a discrete subgroup of the additive group $\mathbb{C}^m$. Then Γ acts by translations on $\mathbb{C}^m$,

$$(k, z) \mapsto t_k(z) := k + z.$$

This action is free and properly discontinuous. For each fixed $k \in \Gamma$, the translation $t_k \colon \mathbb{C}^m \to \mathbb{C}^m$ is biholomorphic. Hence the quotient $T = \Gamma \backslash \mathbb{C}^m$ inherits the structure of a complex manifold such that the covering map $\mathbb{C}^m \to T$ is holomorphic. Now T is diffeomorphic to the $2m$-fold power of a circle, hence T with the above complex structure is called a complex torus. We also note that addition $T \times T \to T$, $(z, w) \mapsto z + w$, is well defined and holomorphic and turns T into a complex Lie group as in Example 8 below.

Let $T = \Gamma \backslash \mathbb{C}^m$ be a complex torus and $f \colon T \to T$ be a biholomorphic map. Then any continuous lift $g \colon \mathbb{C}^m \to \mathbb{C}^m$ of f is Γ-equivariant and biholomorphic. Continuity implies that there is a constant C such that $|g(z)| \leq C(1 + |z|)$ for all $z \in \mathbb{C}^m$. By a standard result from complex analysis, g is affine. Hence f is of the form $f(z) = Az + b$ with $b \in T$ and $A \in \mathrm{Gl}(m, \mathbb{C})$ such that $A(\Gamma) = \Gamma$. Vice versa, for any such $A \in \mathrm{Gl}(m, \mathbb{C})$ and $b \in T$, the map $f \colon T \to T$, $f(z) = Az + b$, is well defined and biholomorphic.

A one-dimensional complex torus is called an *elliptic curve*. It follows from the uniformization theorem that any complex curve diffeomorphic to the torus $S^1 \times S^1$ is an elliptic curve. In particular, the complex structure described in the next example turns $S^1 \times S^1$ into an elliptic curve.

7) *Hopf manifold* (complex structures on $S^{2m-1} \times S^1$). Let $m \geq 1$ and $z \in \mathbb{C}$ be a non-zero complex number of modulus $|z| \neq 1$. Then $\mathbb{Z}$ acts freely and properly discontinuously on $\mathbb{C}^m \setminus \{0\}$ by $(k, v) \mapsto z^k \cdot v$. The quotient $M = (\mathbb{C}^m \setminus \{0\})/\mathbb{Z}$ is called a Hopf manifold. It is an exercise to show that M is diffeomorphic to $S^{2m-1} \times S^1$. (Hint: Consider the case $z = 2$ first.) Since multiplication by z^k is biholomorphic, M inherits from $\mathbb{C}^m \setminus \{0\}$ the structure of a complex manifold.

A generalization of this example is due to Calabi and Eckmann, who showed that the product of odd-dimensional spheres carries a complex structure, see [CE] (Example 2.5 in [KN, Chapter IX]).

8) *Complex Lie groups.* As an open subset of $\mathbb{C}^{n \times n}$, the general linear group $G = \mathrm{Gl}(n, \mathbb{C})$ is a complex manifold of complex dimension n^2, and multiplication $G \times G \to G$ and inversion $G \to G$ are holomorphic maps. The special linear group $\mathrm{Sl}(n, \mathbb{C}) \subset \mathrm{Gl}(n, \mathbb{C})$ is a complex Lie subgroup of complex codimension 1. If V is a complex vector space of dimension n, then any isomorphism $B \colon V \to \mathbb{C}^n$ identifies the general linear group $\mathrm{Gl}(V)$ of V with $\mathrm{Gl}(n, \mathbb{C})$ and turns $\mathrm{Gl}(V)$ into a complex Lie group, independently of the choice of B, and the special linear group $\mathrm{Sl}(V)$ is a complex Lie subgroup of complex codimension 1.

The unitary group $\mathrm{U}(n)$ is not a complex Lie group – recall that its defining equation is not holomorphic. In fact, the Lie algebra $\mathfrak{g}$ of a complex Lie group G is a complex vector space and the adjoint representation Ad of G is a holomorphic map into the complex vector space of complex linear endomorphisms of $\mathfrak{g}$. Hence Ad is constant if G is compact. It follows that compact complex Lie groups are Abelian, that is, complex tori.

For a Lie group G, a *complexification* of G consists of a complex Lie group $G_{\mathbb{C}}$ together with an inclusion $G \to G_{\mathbb{C}}$ such that any smooth morphism $G \to H$, where H is a complex Lie group, extends uniquely to a holomorphic morphism $G_{\mathbb{C}} \to H$. Clearly, if $G_{\mathbb{C}}$ exists, then it is unique up to isomorphism. For example, $\mathrm{Sl}(n, \mathbb{C})$ is the complexification of $\mathrm{Sl}(n, \mathbb{R})$. Any connected compact Lie group has a complexification [Bu, Section 27]; e.g., the complexifications of $\mathrm{SO}(n)$, $\mathrm{SU}(n)$, and $\mathrm{U}(n)$ are $\mathrm{SO}(n, \mathbb{C})$, $\mathrm{Sl}(n, \mathbb{C})$, and $\mathrm{Gl}(n, \mathbb{C})$, respectively.

9) *Homogeneous spaces*. We say that a complex manifold M is *homogeneous* if the group of automorphisms of M is transitive on M. For example, if G is a complex Lie group and H is a closed complex Lie subgroup of G, then the quotient G/H is in a unique way a homogeneous complex manifold such that the natural left action by G on G/H and the projection $G \to G/H$ are holomorphic. *Flag manifolds*, that is, coadjoint orbits of connected compact Lie groups, are homogeneous complex manifolds. In fact, if G is a connected compact Lie group and G_λ is the stabilizer of some $\lambda \in \mathfrak{g}^*$ under the coadjoint representation, then the inclusion of G into its complexification $G_{\mathbb{C}}$ induces an isomorphism $G/G_\lambda \to G_{\mathbb{C}}/P_\lambda$, where P_λ is a suitable parabolic subgroup of $G_{\mathbb{C}}$ associated to λ, a closed complex Lie subgroup of $G_{\mathbb{C}}$, see Section 4.12 in [DK]. For example, $\mathrm{U}(n)/T = \mathrm{Gl}(n, \mathbb{C})/B$, where T is the maximal torus of diagonal matrices in $\mathrm{U}(n)$ and B is the Borel group of all upper triangular matrices in $\mathrm{Gl}(n, \mathbb{C})$, the stabilizer of the standard flag in $\mathbb{C}^n$. For more on flag manifolds, see Chapter 8 in [Bes].

10) *Projective varieties*. A closed subset $V \subset \mathbb{C}P^n$ is called a (complex) projective variety if, locally, V is defined by a set of complex polynomial equations. Outside of its singular locus, that is, away from the subset where the defining equations do not have maximal rank, a projective variety is a complex submanifold of $\mathbb{C}P^n$. We say that V is *smooth* if its singular locus is empty. A well known theorem of Chow says that any closed complex submanifold of $\mathbb{C}P^n$ is a smooth projective variety, see [GH, page 167].

We say that V is a *rational curve* if V is smooth and biholomorphic to $\mathbb{C}P^1$. For example, consider the complex curve $C = \{[z] \in \mathbb{C}P^2 \mid z_0^2 = z_1 z_2\}$ in $\mathbb{C}P^2$, which is contained in $U_1 \cup U_2$. On $U_1 \cap C$ we have $z_2/z_1 = (z_0/z_1)^2$, hence we may use $u_1 = z_0/z_1$ as a holomorphic coordinate for C on $U_1 \cap C$. Similarly, on $U_2 \cap C$ we have $z_1/z_2 = (z_0/z_2)^2$ and we may use $u_2 = z_0/z_2$ as a holomorphic coordinate for C on $U_2 \cap C$. The coordinate transformation on $U_1 \cap U_2 \cap C$ is $u_2 = 1/u_1$. Thus C is biholomorphic to $\mathbb{C}P^1$ and hence is a rational curve in $\mathbb{C}P^2$. In this example, the defining equation has degree two. The map $\mathbb{C}P^1 \ni [1, z] \mapsto [1, z, z^2, \dots, z^m] \in \mathbb{C}P^m$ extends to a holomorphic embedding of $\mathbb{C}P^1$ into $\mathbb{C}P^m$, and the maximal degree of the obvious defining equations of the image is m.

Let M be a complex manifold. We say that a complex vector bundle $E \to M$ is *holomorphic* if E is equipped with a maximal atlas of trivializations whose transition functions are holomorphic. Such an atlas turns E into a complex manifold such that the projection $E \to M$ is holomorphic.

2.3 Examples. 1) The tangent bundle TM together with its complex structure J is a complex vector bundle over M. The usual coordinates for the tangent bundle have holomorphic transition maps and thus turn TM into a complex manifold and holomorphic vector bundle over M.

2) If $E \to M$ is holomorphic, then the complex tensor bundles associated to E are holomorphic. For example, the dual bundle E^* is holomorphic. Note however that $TM \otimes_{\mathbb{R}} \mathbb{C}$ is not a holomorphic vector bundle over M in any natural way.

3) Let $E \to M$ be a holomorphic vector bundle and $f : N \to M$ be a holomorphic map. Then the pull back $f^*E \to N$ is holomorphic.

4) The universal bundle $U_r V \to G_r V$ is holomorphic.

Let (z, U) be a holomorphic coordinate chart of M. As usual, we let $z^j = x^j + iy^j$ and write the corresponding coordinate frame as $(X_1, Y_1, \dots, X_m, Y_m)$. For $p \in U$, we define a complex structure J_p on T_pM by

$$J_p X_j(p) = Y_j(p), \quad J_p Y_j(p) = -X_j(p). \tag{2.4}$$

Since the transition maps of holomorphic coordinate charts are holomorphic, J_p is independent of the choice of holomorphic coordinates. We obtain a smooth field $J = (J_p)$ of complex structures on TM.

Vice versa, if M is a smooth manifold of real dimension $2m$, then a smooth field $J = (J_p)$ of complex structures on TM is called an *almost complex structure* of M. An almost complex structure $J = J_p$ is called a *complex structure* if it comes from a complex structure on M as in (2.4) above. Any almost complex structure on a surface is a complex structure (existence of isothermal coordinates). A celebrated theorem of Newlander and Nirenberg [NN] says that an almost complex structure is a complex structure if and only if its *Nijenhuis tensor* or *torsion* N vanishes, where, for vector fields X and Y on M,

$$N(X, Y) = 2\{[JX, JY] - [X, Y] - J[X, JY] - J[JX, Y]\}. \tag{2.5}$$

For an instructive discussion of the Newlander–Nirenberg theorem and its proof, see [Ka2, Section 6.3].

2.6 Exercises. 1) N is a tensor. Compare also Exercises 2.15 and 2.32.

2) Let ∇ be a torsion free connection and J be an almost complex structure on M. Show that

$$\frac{1}{2}N(X, Y) = \nabla J(JX, Y) - J\nabla J(X, Y) - \nabla J(JY, X) + J\nabla J(Y, X)$$

to conclude that J is a complex structure if $\nabla J = 0$.

2.7 Example. In the normed division algebra Ca of Cayley numbers [4], consider the sphere S^6 of purely imaginary Cayley numbers of norm one. For a point $p \in S^6$ and tangent vector $v \in T_pS^6$, define

$$J_p v := p \cdot v,$$

[4] Chapter 15 in [Ad] is a good reference to Cayley numbers.

where the dot refers to multiplication in Ca. Since p is purely imaginary of length one, $p \cdot (p \cdot x) = -x$ for all $x \in \mathrm{Ca}$. It follows that $J_p v \in T_p S^6$ and that

$$J_p^2 v = p \cdot (p \cdot v) = -v.$$

Hence $J = (J_p)$ is an almost complex structure on S^6. Since $|x \cdot y| = |x| \cdot |y|$ for all $x, y \in \mathrm{Ca}$ and $p \in S^6$ has norm one, J is norm preserving.

By parallel translation along great circle arcs through p, extend v to a vector field V in a neighborhood of p in S^6. Then V is parallel at p. Along the great circle arc $\cos t \cdot p + \sin t \cdot u$ in the direction of a unit vector u in $T_p S^6$, the vector field JV is given by $(\cos t \cdot p + \sin t \cdot u) \cdot V(\cos t \cdot p + \sin t \cdot u)$. Hence

$$dJV(u) = u \cdot v + p \cdot dV(u) = u \cdot v + p \cdot S(u, v),$$

where S denotes the second fundamental form of S^6. For $v, w \in T_p S^6$, we get

$$N(v, w) = 2\{(p \cdot v) \cdot w - (p \cdot w) \cdot v - p \cdot (v \cdot w) + p \cdot (w \cdot v)\} = 4[p, v, w],$$

where $[x, y, z] = (x \cdot y) \cdot z - x \cdot (y \cdot z)$ denotes the associator of $x, y, z \in \mathrm{Ca}$. We conclude that $N \neq 0$ and hence that J does not come from a complex structure on S^6. It is a famous open problem whether S^6 carries any complex structure.

2.8 Exercise. View the sphere S^2 as the space of purely imaginary quaternions of norm one and discuss the corresponding almost complex structure on S^2.

2.1 Complex Vector Fields. Let V be a real vector space, and let J be a complex structure on V. We extend J complex linearly to the complexification $V_{\mathbb{C}} = V \otimes_{\mathbb{R}} \mathbb{C}$ of V,

$$J(v \otimes \alpha) := (Jv) \otimes \alpha. \tag{2.9}$$

Then we still have $J^2 = -1$, hence $V_{\mathbb{C}}$ is the sum

$$V_{\mathbb{C}} = V' \oplus V'' \tag{2.10}$$

of the eigenspaces V' and V'' for the eigenvalues i and $-i$, respectively. The maps

$$V \to V', \ v \mapsto v' := \frac{1}{2}(v - iJv), \quad V \to V'', \ v \mapsto v'' := \frac{1}{2}(v + iJv), \tag{2.11}$$

are complex linear and conjugate linear isomorphisms, respectively, if we consider V together with J as a complex vector space.

Let M be a smooth manifold with an almost complex structure J and $T_{\mathbb{C}}M = TM \otimes_{\mathbb{R}} \mathbb{C}$ be the complexified tangent bundle. As in (2.10), we have the eigenspace decomposition with respect to J,

$$T_{\mathbb{C}}M = T'M \oplus T''M. \tag{2.12}$$

The decomposition in (2.11) shows that $T'M$ and $T''M$ are smooth subbundles of the complexified tangent bundle $T_{\mathbb{C}}M$. Moreover, by (2.11) the maps

$$TM \to T'M, \ v \mapsto v', \quad \text{and} \quad TM \to T''M, \ v \mapsto v'', \tag{2.13}$$

are complex linear respectively conjugate linear isomorphisms of complex vector bundles over M, where multiplication by i on TM is given by J.

A *complex vector field* of M is a section of $T_{\mathbb{C}}M$. Any such field can be written in the form $Z = X + iY$, where X and Y are vector fields of M, that is, sections of TM. Complex vector fields act as complex linear derivations on smooth complex valued functions. We extend the Lie bracket complex linearly to complex vector fields,

$$[X + iY, U + iV] := [X, U] - [Y, V] + i([X, V] + [Y, U]). \tag{2.14}$$

2.15 Exercise. The Nijenhuis tensor associated to J vanishes iff $T'M$ is an involutive distribution of $T_{\mathbb{C}}M$, that is, if $[Z_1, Z_2]$ is a section of $T'M$ whenever Z_1 and Z_2 are. And similarly for $T''M$. More precisely,

$$N(X, Y)'' = -8[X', Y']'' \quad \text{and} \quad N(X, Y)' = -8[X'', Y'']',$$

where X and Y are vector fields on M.

Suppose from now on that M is a complex manifold. Then TM (with complex multiplication defined via J) is a holomorphic vector bundle over M. The isomorphism $TM \ni v \mapsto v' \in T'M$ as in (2.13) turns $T'M$ into a holomorphic vector bundle. The bundle $T''M$ is a smooth complex vector bundle over M, but not a holomorphic vector bundle in a natural way.

Let (z, U) be a holomorphic coordinate chart for M. Write $z^j = x^j + iy^j$ and set

$$Z_j := \frac{\partial}{\partial z^j} = \frac{1}{2}(X_j - iY_j), \quad \bar{Z}_j := \frac{\partial}{\partial \bar{z}^j} = \frac{1}{2}(X_j + iY_j), \tag{2.16}$$

where $(X_1, Y_1, \dots, X_m, Y_m)$ is the coordinate frame associated to the coordinates $(x^1, y^1, \dots, x^m, y^m)$. In the notation of (2.13), $Z_j = X_j'$ and $\bar{Z}_j = X_j''$. Similarly, any complex vector field Z has components

$$Z' = \frac{1}{2}(Z - iJZ) \quad \text{and} \quad Z'' = \frac{1}{2}(Z + iJZ) \tag{2.17}$$

in $T'M$ and $T''M$, respectively. We note that $(X_1, \dots, X_m)$ is a local holomorphic frame for TM considered as a holomorphic vector bundle over M and that $(Z_1, \dots, Z_m)$ is the corresponding local holomorphic frame of $T'M$.

We say that a real vector field X on M is *automorphic* if the flow of X preserves the complex structure J of M. This holds iff the Lie derivative $L_X J = 0$ or, equivalently, iff $[X, JY] = J[X, Y]$ for all vector fields Y on M. The space $\mathfrak{a}(M)$ of automorphic vector fields on M is a Lie algebra with respect to the Lie bracket of vector fields.

2.18 Proposition. *A real vector field X on M is automorphic iff it is holomorphic as a section of the holomorphic vector bundle TM. The complex structure J turns $\mathfrak{a}(M)$ into a complex Lie algebra.*

Proof. To be automorphic or holomorphic is a local property. Hence we can check the equivalence of the two properties in holomorphic coordinates (z, U). Then the vector field is given by a smooth map $X: U \to \mathbb{C}^m$ and J is given by multiplication by i.

Let $Y : U \to \mathbb{C}^m$ be another vector field. Then the Lie bracket of X with Y is given by $dX(Y) - dY(X)$. Hence we get

$$\begin{aligned}[X, iY] &= dX(iY) - idY(X)\\ &= \partial X(iY) + \bar\partial X(iY) - idY(X)\\ &= i\partial X(Y) - i\bar\partial X(Y) - idY(X)\\ &= i\partial X(Y) + i\bar\partial X(Y) - idY(X) - 2i\bar\partial X(Y)\\ &= i[X, Y] - 2i\bar\partial X(Y).\end{aligned}$$

Hence $[X, iY] = i[X, Y]$ iff $\bar\partial X(Y) = 0$. □

2.19 Proposition. *If M is a closed complex manifold. then* $\dim \mathfrak{a}(M) < \infty$.

Proof. The space of holomorphic sections of a holomorphic vector bundle E over M is precisely the kernel of the elliptic differential operator $\bar\partial$ on the space of smooth sections of E, compare (3.4). □

2.20 Remark. A celebrated theorem of Bochner and Montgomery states that, for a closed complex manifold M, the group $\mathrm{Aut}(M, J)$ of automorphisms $f : M \to M$ is a complex Lie group with respect to the compact-open topology[5] and that $\mathfrak{a}(M)$ is the Lie algebra of $\mathrm{Aut}(M, J)$, see [BM]. In particular, $\mathrm{Aut}(M, J)$ is either trivial, or a complex torus, or is not compact, compare Example 2.2.8.

2.2 Differential Forms. As above, let V be a real vector space with complex structure J. The decomposition of $V_\mathbb{C}$ in (2.10) determines a decomposition of the space of complex valued alternating forms on $V_\mathbb{C}$,

$$\begin{aligned}\Lambda^r V_\mathbb{C}^* &= \Lambda^r(V' \oplus V'')^*\\ &= \oplus_{p+q=r}\big(\Lambda^p(V')^* \otimes \Lambda^q(V'')^*\big) =: \oplus_{p+q=r}\Lambda^{p,q}V_\mathbb{C}^*.\end{aligned} \tag{2.21}$$

Alternating forms on $V_\mathbb{C}$ correspond to complex multilinear extensions of complex valued alternating forms on V. In this interpretation, $\Lambda^{p,q}V_\mathbb{C}^*$ corresponds to complex valued alternating r-forms φ on V, $r = p + q$, such that, for $\alpha \in \mathbb{C}$,

$$\varphi(\alpha v_1, \dots, \alpha v_r) = \alpha^p\bar\alpha^q\varphi(v_1, \dots, v_r), \tag{2.22}$$

[5] This is a subtle point. Without the assertion about the topology, the theorem is a trivial consequence of Proposition 2.19. There is a corresponding common misunderstanding in the case of isometry groups of Riemannian manifolds and of other groups of automorphisms.

where we view V together with J as a complex vector space as usual. To see this, write $v_j = v_j' + v_j''$ as in (2.11). We call elements of $\Lambda^{p,q}V_{\mathbb{C}}^*$ alternating forms on V of type (p, q). An alternating r-form φ of type (p, q) satisfies

$$(J^*\varphi)(v_1, \dots, v_r) := \varphi(Jv_1, \dots, Jv_r) = i^{p-q}\varphi(v_1, \dots, v_r), \tag{2.23}$$

but this does not characterize the type. We note however that a non-zero alternating r-form φ satisfies $J^*\varphi = \varphi$ iff r is even and φ is of type $(r/2, r/2)$.

Conjugation maps $\Lambda^{p,q}V_{\mathbb{C}}^*$ to $\Lambda^{q,p}V_{\mathbb{C}}^*$, and $\Lambda^{p,p}V_{\mathbb{C}}^*$ is invariant under conjugation. The space of complex (p, p)-forms fixed under conjugation is the space of real valued forms of type (p, p).

Suppose now that M is a complex manifold with complex structure J. Note that $A^*(M, \mathbb{C}) = \Lambda^* T_{\mathbb{C}}^*M$, where $T_{\mathbb{C}}^*M = T^*M \otimes_{\mathbb{R}} \mathbb{C}$ is the complexified cotangent bundle. As in (2.21), we have the decomposition

$$A^r(M, \mathbb{C}) = \oplus_{p+q=r} A^{p,q}(M, \mathbb{C}), \tag{2.24}$$

where $A^{p,q}(M, \mathbb{C}) := \Lambda^{p,q}T_{\mathbb{C}}^*M$. The complex line bundle $K_M := A^{m,0}(M, \mathbb{C})$ plays a special role, it is called the *canonical bundle*. Smooth sections of $A^{p,q}(M, \mathbb{C})$ are called *differential forms* of type (p, q), the space of such differential forms is denoted $\mathcal{A}^{p,q}(M, \mathbb{C})$.

Let $z\colon U \to U'$ be a holomorphic coordinate chart for M. Write $z^j = x^j + iy^j$ and set

$$dz^j = dx^j + i dy^j \quad \text{and} \quad d\bar{z}^j = dx^j - i dy^j, \tag{2.25}$$

differential forms of type $(1, 0)$ and $(0, 1)$, respectively. In terms of these, a differential form φ of type (p, q) is given by a linear combination

$$\sum_{J,K} a_{JK}\, dz^J \wedge d\bar{z}^K = \sum_{J,K} a_{JK}\, dz^{i_1} \wedge \dots \wedge dz^{i_p} \wedge d\bar{z}^{j_1} \wedge \dots \wedge d\bar{z}^{j_q}, \tag{2.26}$$

where J and K run over multi-indices $j_1 < \dots < j_p$ and $k_1 < \dots < k_q$. Under a transformation $z \circ w^{-1}$ of holomorphic coordinates of M, we have

$$dz^j = \sum \frac{\partial z^j}{\partial w^k} dw^k \quad \text{and} \quad d\bar{z}^j = \sum \frac{\partial \bar{z}^j}{\partial \bar{w}^k} d\bar{w}^k. \tag{2.27}$$

This shows that the natural trivializations by the forms dz^I as above turn the bundles $A^{p,0}(M, \mathbb{C})$ into holomorphic bundles over M. We also see that the complex vector bundles $A^r(M, \mathbb{C})$ and $A^{p,q}(M, \mathbb{C})$ are not holomorphic in any natural way for $0 < r \le 2m$ and $0 < q \le m$, respectively.

A quick computation gives

$$\begin{aligned} d\varphi &= \sum \big(X_j(a_{JK})dx^j + Y_j(a_{JK})dy^j\big) \wedge dz^J \wedge d\bar{z}^K \\ &= \sum \big(Z_j(a_{JK})dz^j + \bar{Z}_j(a_{JK})d\bar{z}^j\big) \wedge dz^J \wedge d\bar{z}^K \\ &=: \partial\varphi + \bar{\partial}\varphi. \end{aligned} \tag{2.28}$$

The type of $\partial\varphi$ is $(p+1, q)$, the type of $\overline{\partial}\varphi$ is $(p, q+1)$, hence they are well defined, independently of the choice of holomorphic coordinates. Now $d = \partial + \overline{\partial}$ and $d^2 = 0$. Hence by comparing types, we get

$$\partial^2 = 0, \quad \overline{\partial}^2 = 0, \quad \partial\overline{\partial} = -\overline{\partial}\partial. \tag{2.29}$$

In particular, we get differential cochain complexes

$$\cdots \xrightarrow{\overline{\partial}} \mathcal{A}^{p,q-1}(M,\mathbb{C}) \xrightarrow{\overline{\partial}} \mathcal{A}^{p,q}(M,\mathbb{C}) \xrightarrow{\overline{\partial}} \mathcal{A}^{p,q+1}(M,\mathbb{C}) \xrightarrow{\overline{\partial}} \cdots \tag{2.30}$$

whose cohomology groups $H^{p,q}(M,\mathbb{C})$ are called *Dolbeault cohomology groups* of M. Their dimensions, $h^{p,q}(M,\mathbb{C}) = \dim_{\mathbb{C}} H^{p,q}(M,\mathbb{C})$, are called *Hodge numbers* of M. They are invariants associated to the complex structure of M.

The kernel $\Omega^p(M)$ of $\overline{\partial}$ on $\mathcal{A}^{p,0}(M,\mathbb{C})$ consists precisely of the holomorphic sections of the holomorphic vector bundle $A^{p,0}(M,\mathbb{C})$. These are called *holomorphic forms* of degree p. By definition, $\Omega^p(M) \cong H^{p,0}(M,\mathbb{C})$. The alternating sum

$$\chi(M,\mathcal{O}) := \sum (-1)^p h^{p,0}(M,\mathbb{C}) = \sum (-1)^p \dim_{\mathbb{C}} \Omega^p(M) \tag{2.31}$$

is called the *arithmetic genus* of M.

2.32 Exercise. Let M be a smooth manifold with an almost complex structure J. As in the case of complex manifolds, we have the decomposition

$$A^r(M,\mathbb{C}) = \oplus_{p+q=r} A^{p,q}(M,\mathbb{C})$$

into types. If φ is a smooth complex valued function, that is $\varphi \in \mathcal{A}^0(M,\mathbb{C})$, then we can decompose as in the case of complex manifolds,

$$d\varphi = d^{1,0}\varphi + d^{0,1}\varphi \in \mathcal{A}^{1,0}(M,\mathbb{C}) + \mathcal{A}^{0,1}(M,\mathbb{C}),$$

where $d^{1,0} = \partial$ and $d^{0,1} = \overline{\partial}$ if J is a complex structure. However, if φ is of type $(1,0)$, then a new term may arise,

$$d\varphi = d^{1,0}\varphi + d^{0,1}\varphi + d^{-1,2}\varphi \in \mathcal{A}^{2,0}(M,\mathbb{C}) + \mathcal{A}^{1,1}(M,\mathbb{C}) + \mathcal{A}^{0,2}(M,\mathbb{C}),$$

and similarly for $\varphi \in \mathcal{A}^{0,1}(M,\mathbb{C})$. Show that $d^{-1,2}$ is a tensor field and that

$$\varphi(N(X,Y)) = 8 \cdot d^{-1,2}\varphi(X,Y)$$

for all $\varphi \in \mathcal{A}^{1,0}(M,\mathbb{C})$ and vector fields X, Y on M. Any differential form is locally a finite sum of decomposable differential forms. Conclude that

$$d\varphi \in \mathcal{A}^{p+2,q-1}(M,\mathbb{C}) + \mathcal{A}^{p+1,q}(M,\mathbb{C}) + \mathcal{A}^{p,q+1}(M,\mathbb{C}) + \mathcal{A}^{p-1,q+2}(M,\mathbb{C})$$

for any differential form φ of type (p,q). Which parts of $d\varphi$ are tensorial in φ?

2.3 Compatible Metrics. Let M be a complex manifold with corresponding complex structure J. We say that a Riemannian metric $g = \langle \cdot, \cdot \rangle$ is *compatible* with J if

$$\langle JX, JY \rangle = \langle X, Y \rangle \tag{2.33}$$

for all vector fields X, Y on M. A complex manifold together with a compatible Riemannian metric is called a *Hermitian manifold*[6].

Let M be a complex manifold as above. If g is a compatible Riemannian metric on M, then the complex bilinear extension of g to $T_{\mathbb{C}}M$, also denoted g or $\langle \cdot, \cdot \rangle$, is symmetric and satisfies the following three conditions:

$$\begin{aligned} \langle \bar{Z}_1, \bar{Z}_2 \rangle &= \overline{\langle Z_1, Z_2 \rangle}; & & \\ \langle Z_1, Z_2 \rangle &= 0 & &\text{for } Z_1, Z_2 \text{ in } T'M; \\ \langle \bar{Z}, Z \rangle &> 0 & &\text{unless } Z = 0. \end{aligned} \tag{2.34}$$

Vice versa, a symmetric complex bilinear form $\langle \cdot, \cdot \rangle$ on $T_{\mathbb{C}}M$ satisfying these three conditions is the complex bilinear extension of a Riemannian metric satisfying (2.33). Then (2.33) also holds for the complex linear extension of J to $T_{\mathbb{C}}M$ and complex vector fields X and Y.

2.35 Proposition. *Let M be a complex manifold with complex structure J. Then a Riemannian metric $g = \langle \cdot, \cdot \rangle$ is compatible with J iff about each point $p_0 \in M$, there are holomorphic coordinates*

$$z = (z^1, \dots, z^m) = (x^1 + iy^1, \dots, x^m + iy^m)$$

such that the associated coordinate frame $(X_1, Y_1, \dots, X_m, Y_m)$ satisfies

$$\langle X_j, X_k \rangle (p_0) = \langle Y_j, Y_k \rangle (p_0) = \delta_{jk} \quad \textit{and} \quad \langle X_j, Y_k \rangle (p_0) = 0.$$

The proof of Proposition 2.35 is straightforward and left as an exercise. We note the following immediate consequence.

2.36 Corollary. *Let M be a Hermitian manifold. Then the type decomposition*

$$A^r(M, \mathbb{C}) = \oplus_{p+q=r} A^{p,q}(M, \mathbb{C})$$

is orthogonal with respect to the induced Hermitian metric $(\varphi, \psi) = \langle \bar{\varphi}, \psi \rangle$. In particular, $*\varphi$ has type $(m-q, m-p)$ if φ has type (p, q).

Proof. The first assertion is clear from Proposition 2.35. The second assertion follows from the first since the volume form has type (m, m)[7]. □

[6] The usual arguments give the existence of compatible Riemannian metrics.

[7] Recall that $dz \wedge d\bar{z} = -2i dx \wedge dy$.

2.4 Blowing Up Points. Consider the universal bundle $U_1\mathbb{C}^m$ over $\mathbb{C}P^{m-1}$. The restriction of the map $U_1\mathbb{C}^m \to \mathbb{C}^m$, $(L, z) \mapsto z$, to the open subset $\{(L, z) \in U_1\mathbb{C}^m \mid z \neq 0\}$ is biholomorphic onto $\mathbb{C}^m \setminus \{0\}$. Thus we can think of $U_1\mathbb{C}^m$ as $\mathbb{C}^m$, where the point 0 is replaced by the set $\{(L, 0) \in U_1\mathbb{C}^m\}$. We identify the latter with $\mathbb{C}P^{m-1}$ and thus have *blown up* the point 0 of $\mathbb{C}^m$ to $\mathbb{C}P^{m-1}$. A similar construction can be carried out for points in complex manifolds.

2.37 Remark. For any $r > 0$, we can identify $S_r = \{(L, z) \in U_1\mathbb{C}^m \mid |z| = r\}$ with the sphere of radius r in $\mathbb{C}^m$. Then the projection $\pi : S_r \to \mathbb{C}P^{m-1}$ turns into the Hopf fibration: The fibers of π intersect S_r in Hopf circles, that is, the intersections of complex lines in $\mathbb{C}^m$ with the sphere S_r. Renormalizing the given Riemannian metric on S_r (of sectional curvature $1/r^2$) by adding the pull back of any fixed Riemannian metric g on $\mathbb{C}P^{m-1}$, we obtain a Riemannian metric g_r on the sphere S^{2m-1}. We can think of (S^{2m-1}, g_r) as a collapsing family of Riemannian manifolds with limit $(\mathbb{C}P^{m-1}, g)$, as $r \to 0$. The differential geometric significance of this kind of collapse was first recognized by Berger: The sectional curvature of the collapsing spheres stays uniformly bounded as $r \to 0$, see Example 3.35 in [CE], [Kar, page 221], and [CGr].

Let M be a complex manifold and p be a point in M. The blow up of M at p replaces p by the space of complex lines in T_pM. The precise construction goes as follows. Let $z = (z_1, \dots, z_m) \colon U \to U'$ be holomorphic coordinates about p with $z(p) = 0$. In $U' \times \mathbb{C}P^{m-1}$, consider the set

$$V = \{(z, [w]) \mid z^i w^j = z^j w^i \text{ for all } i, j\},$$

where points in $\mathbb{C}P^{m-1}$ are given by their homogeneous coordinates, denoted $[w]$. On the subset $\{w^j \neq 0\}$ of $\mathbb{C}P^{m-1}$, V is defined by the $m - 1$ independent equations

$$z^i = \frac{w^i}{w^j} z^j, \quad i \neq j,$$

the other equations follow. Hence the system of equations defining V has constant rank $m - 1$, and hence V is a complex submanifold of $U' \times \mathbb{C}P^{m-1}$ of dimension m with

$$S = V \cap \{z = 0\} \cong \mathbb{C}P^{m-1}.$$

On $V \setminus S$, $[w]$ is determined by z, hence

$$V \setminus S \cong U \setminus \{p\}.$$

More precisely, the canonical map

$$V \setminus S \to U' \setminus \{0\}, \quad (z, [w]) \mapsto z,$$

is biholomorphic. We use it to glue V to $M \setminus \{p\}$ and obtain a complex manifold $\tilde{M}$, the blow up of M at p, together with a holomorphic map

$$\pi : \tilde{M} \to M$$

such that $\pi^{-1}(p) = S$ and $\pi : \tilde{M} \setminus S \to M \setminus \{p\}$ is biholomorphic.

We consider V as an open subset of $\tilde{M}$. Choose $\varepsilon > 0$ such that the image U' of the holomorphic coordinates z contains the ball of radius $\varepsilon > 0$ about 0 in $\mathbb{C}^m$. Then the map

$$V_\varepsilon := \{(z, [w]) \in V \mid |z| < \varepsilon\} \to \{(L, z) \in U_1\mathbb{C}^m \mid |z| < \varepsilon\},$$
$$(z, [w]) \mapsto ([w], z),$$

is biholomorphic. Hence a neighborhood of S in $\tilde{M}$ is biholomorphic to a neighborhood of the zero-section of $U_1\mathbb{C}^m$ such that the map $\sigma : V_\varepsilon \to \mathbb{C}P^{m-1}$, $\sigma(z, [w]) = [w]$, corresponds to the projection. For any $r \in (0, \varepsilon)$, the set

$$S_r = \{(z, [w]) \in V \mid |z| = r\}$$

corresponds to the sphere of radius r in $\mathbb{C}^m$. The fibers of π intersect S_r in Hopf circles, that is, the intersections of complex lines in $\mathbb{C}^m$ with S_r. Thus we can again think of $\pi : S_r \to \mathbb{C}P^{m-1}$ as the Hopf fibration and of the convergence $S_r \to S$, as $r \to 0$, as the collapse of S^{2m-1} to $\mathbb{C}P^{m-1}$ along Hopf circles; compare with Remark 2.37 above.

2.38 Exercises. Let $\tilde{M}$ be the blow up of M at p as above.

1) Let $\hat{z} = f(z)$ be other holomorphic coordinates of M about p with $\hat{z}(p) = 0$, and let $\hat{M}$ be the blow up of M at p with respect to the coordinates $\hat{z}$. Write $f(z) = \sum f_j(z)z^j$, where the maps f_j are holomorphic with $f_j(0) = \partial_j f(0)$. Show that

$$f(z, [w]) = \big(f(z), \big[\sum f_j(z)w^j\big]\big)$$

extends the identity on $M \setminus \{p\}$ to a biholomorphic map $\tilde{M} \to \hat{M}$. In this sense, the blow up of M at p does not depend on the choice of centered holomorphic coordinates.

2) Let $\tilde{f} : \tilde{M} \to N$ be a holomorphic map, where N is another complex manifold. If f is constant on S, then there is a holomorphic map $f : M \to N$ such that $\tilde{f} = f \circ \pi$.

2.39 Remark. By replacing centered holomorphic coordinates about p by a tubular neighborhood of the zero section in the normal bundle, there is a rather immediate generalization of the blow up of points to a blow up of complex submanifolds, see e.g. Section 4.6 in [GH] or Section 2.5 in [Hu].

2.40 Examples. 1) *Kummer surface.* In this example it will be convenient to enumerate coordinates by subindices. Consider the quotient $Q = \mathbb{C}^2/\mathbb{Z}_2$, where $\mathbb{Z}_2 = \{1, -1\}$ acts by scalar multiplication on $\mathbb{C}^2$. Let q_0 be the image of 0 under the natural projection $\mathbb{C}^2 \to Q$. Note that away from 0 and q_0, respectively, the projection is a twofold covering with holomorphic covering transformations, turning $Q \setminus \{q_0\}$ into a complex manifold. Via

$$z_0 = t_1 t_2, \quad z_1 = t_1^2, \quad z_2 = t_2^2,$$

we can identify Q with the algebraic hypersurface

$$H = \{(z_0, z_1, z_2) \in \mathbb{C}^3 \mid z_0^2 = z_1 z_2\} \subset \mathbb{C}^3,$$

which has a singularity at the origin 0. We blow up $\mathbb{C}^3$ at 0 to get

$$\tilde{\mathbb{C}}^3 = \{(z,[w]) \in \mathbb{C}^3 \times \mathbb{C}P^2 \mid z_i w_j = z_j w_i\}.$$

We have

$$\pi^{-1}(H \setminus \{0\}) = \{(z,[w]) \in \tilde{\mathbb{C}}^3 \mid z \neq 0,\, w_0^2 = w_1 w_2\}.$$

In particular, the closure of $\pi^{-1}(H \setminus \{0\})$ in $\tilde{\mathbb{C}}^3$ is the regular hypersurface

$$\tilde{H} = \{(z,[w]) \in \tilde{\mathbb{C}}^3 \mid w_0^2 = w_1 w_2\} \subset \tilde{\mathbb{C}}^3.$$

Thus by blowing up $0 \in \mathbb{C}^3$, we resolved the singularity of H.

Recall that $C = \{[w] \in \mathbb{C}P^2 \mid w_0^2 = w_1 w_2\}$ is an embedded $\mathbb{C}P^1$, compare Example 2.2.10. There is a natural projection $\tilde{H} \to C$. We choose (u_j, z_j) as holomorphic coordinates for $\tilde{H}$ over the preimage of U_j under this projection. The coordinate transformation over the preimage of $U_1 \cap U_2$ is $(u_2, z_2) = (u_1^{-1}, u_1^2 z_1)$. The holomorphic cotangent bundle $A^{1,0}(C,\mathbb{C})$ has du_j as a nowhere vanishing section over U_j, and $du_2 = -u_1^{-2} du_1$ over $U_1 \cap U_2$. We conclude that the identification $(u_1, z_1 du_1) \leftrightarrow (u_1, z_1)$ over U_1 and $(u_2, -z_2 du_2) \leftrightarrow (u_2, z_2)$ over U_2 establishes a biholomorphic map between $\tilde{H}$ and $A^{1,0}(C,\mathbb{C}) = A^{1,0}(\mathbb{C}P^1,\mathbb{C})$.

Let $T^4 = \mathbb{Z}^4 \backslash \mathbb{C}^2$. Scalar multiplication by $\mathbb{Z}_2$ on $\mathbb{C}^2$ descends to an action of $\mathbb{Z}_2$ on T^4. This action has 16 fixed points, namely the points with integral or half-integral coordinates in T^4. At each fixed point $x \in T^4$, the action is locally of the form $\pm 1 \cdot (x+t) = x \pm t$ as above. Thus we can resolve each of the quotient singularities on $T^4/\mathbb{Z}_2$ by the above construction and obtain a compact complex surface, the *Kummer surface*. For more information on the Kummer surface we refer to [Jo, Section 7.3].

2) *Dependence of blow-up on points.*[8] Let $m \geq 2$ and $T = \Gamma \backslash \mathbb{C}^m$ be a complex torus. Let $(p_1, \dots, p_k)$ and $(p_1', \dots, p_l')$ be two tuples of pairwise different points in T. Let M and M' be the blow ups of T in the points $p_1, \dots, p_k$ and $p_1', \dots, p_l'$, respectively, and let $\pi\colon M \to T$ and $\pi'\colon M' \to T$ be the projections. Let S_i and S_j' be the preimages of p_i and p_j' under π and π', respectively.

Let $f\colon M \to M'$ be a holomorphic map. Since $S_i \cong \mathbb{C}P^{m-1}$ is simply connected, the restriction of $\pi' \circ f$ to S_i lifts to a holomorphic map $S_i \to \mathbb{C}^m$. Now S_i is a closed complex manifold, hence any such lift is constant. It follows that $\pi' \circ f$ maps S_i to a point in T. In particular, there is a holomorphic map $g\colon T \to T$ such that $\pi' \circ f = g \circ \pi$.

Suppose now that f is biholomorphic. Then f is not constant on S_i (we assume $m \geq 2$). By what we said above, it follows that f maps each S_i biholomorphically to an S_j'. Thus $k = l$ and, up to renumeration, $f(S_i) = S_i'$ for all i. Moreover, the induced map $g\colon T \to T$ is biholomorphic with $g(p_i) = p_i'$ for all i.

Let now $k = l = 2$, $p_1 = p_1' = 0$, $p_2 = p$, $p_2' = p'$. In Example 2.2.6 above we showed that any biholomorphic map of T is of the form $h(z) = Az + b$ with $b \in T$ and $A \in \mathrm{Gl}(m,\mathbb{C})$ such that $A(\Gamma) = \Gamma$. Hence the above g is of the form $g(z) = Az$

[8] I owe this example to Daniel Huybrechts.

for some $A \in \mathrm{Gl}(m, \mathbb{C})$ with $A(\Gamma) = \Gamma$ and $Ap = p'$ (in T). On the other hand, there are pairs of points $p, p' \in T \setminus \{0\}$ such that there is no $A \in \mathrm{Gl}(m, \mathbb{C})$ with $A(\Gamma) = \Gamma$ and $Ap = p'$. Then by what we just said, the corresponding blow ups M and M' of T in 0, p and 0, p', respectively, are not biholomorphic.

2.41 Exercise. In terms of oriented smooth manifolds, the blow up $\tilde{M}$ of M at $p \in M$ corresponds to the connected sum $M \# \overline{\mathbb{C}P}^m$, where $\overline{\mathbb{C}P}^m$ denotes $\mathbb{C}P^m$ with orientation opposite to the standard one: Choose w/w_0 as a coordinate about $q = [1, 0, \dots, 0]$ in $U_0 = \{[w_0, w] \mid w_0 \neq 0\} \subset \mathbb{C}P^m$. Let $\varepsilon > 0$ and $V_\varepsilon \subset \tilde{M}$ be as above. Set

$$U'_\varepsilon = \{[w_0, w] \in \overline{\mathbb{C}P}^m \mid |w_0| < \varepsilon |w|\}.$$

Use coordinates z about p as in the definition of blow ups and w/w_0 as above to define the connected sum $M \# \overline{\mathbb{C}P}^m$. Then the map $f\colon M \# \overline{\mathbb{C}P}^m \to \tilde{M}$,

$$f(p') = \begin{cases} p' & \text{if } p' \in M \setminus \{p\}, \\ (w_0 w/|w|^2, [w]) \in V_\varepsilon & \text{if } p' = [w_0, w] \in U'_\varepsilon, \end{cases}$$

is well defined and an orientation preserving diffeomorphism. In particular, as a smooth manifold, the blow up $\tilde{M}$ does not depend on the choice of $p \in M$.

We conclude our discussion of blow ups with a fact on the automorphism group of a compact complex surface which we cannot prove in the framework of these lecture notes, but which will provide us with an important example in Subsection 7.4. For a closed complex manifold M with complex structure J, denote by $\mathrm{Aut}(M) = \mathrm{Aut}(M, J)$ the group of biholomorphic transformations of M, endowed with the compact-open topology, and by $\mathrm{Aut}_0(M)$ the component of the identity in $\mathrm{Aut}(M)$.

2.42 Proposition. *Let M be a compact complex surface and $\pi : M_p \to M$ be the blow up of M in $p \in M$. Then automorphisms in* $\mathrm{Aut}_0(M_p)$ *leave the* exceptional divisor *$S = \pi^{-1}(p)$ invariant and, via restriction to $M \setminus S$,*

$$\mathrm{Aut}_0(M_p) \cong \{\Phi \in \mathrm{Aut}_0(M) \mid \Phi(p) = p\} =: \mathrm{Aut}_0(M, p).$$

The action of $\Phi \in$ $\mathrm{Aut}_0(M, p)$ *on $S \cong P(T_pM)$ is induced by $d\Phi_p$.*

The only issue is to show that an automorphism of M_p in $\mathrm{Aut}_0(M_p)$ leaves S invariant. Given some facts about topological intersection properties of analytic cycles, the proof of this is actually quite simple and geometric, see [GH], Chapter 4.1. A corresponding statement would be wrong for the full automorphism group, since if M itself is already a blow up, there may be automorphisms permuting the various exceptional divisors.

2.43 Example. Consider the blow up of $\mathbb{C}P^2$ in one or two points, which we choose to be [1, 0, 0] and [1, 0, 0], [0, 1, 0], respectively. By Proposition 2.42, the respective automorphism groups are

$$\left\{ \begin{pmatrix} 1 & * & * \\ 0 & * & * \\ 0 & * & * \end{pmatrix} \in \mathrm{Gl}(3, \mathbb{C}) \right\} \quad \text{and} \quad \left\{ \begin{pmatrix} 1 & 0 & * \\ 0 & * & * \\ 0 & 0 & * \end{pmatrix} \in \mathrm{Gl}(3, \mathbb{C}) \right\}. \tag{2.44}$$

For a blow up of $\mathbb{C}P^2$ in three points, the automorphism group, and hence the complex structure, clearly depends on the choice of points, i.e., on whether they are in general position or not.

3 Holomorphic Vector Bundles

Let $E \to M$ be a holomorphic vector bundle. For $U \subset M$ open, we denote by $\mathcal{O}(U, E)$ the space of holomorphic sections of E over U, a module over the ring $\mathcal{O}(U)$. If $(\Phi_1, \dots, \Phi_k)$ is a holomorphic frame of E over U, then

$$\mathcal{O}(U)^k \ni \varphi \mapsto \varphi^\mu \Phi_\mu \in \mathcal{O}(U, E) \tag{3.1}$$

is an isomorphism.

3.1 Dolbeault Cohomology. We now consider differential forms on M with values in E. Since M is complex, we can distinguish forms according to their type as before,

$$A^r(M, E) = \sum_{p+q=r} A^{p,q}(M, E), \tag{3.2}$$

where $A^{p,q}(M, E) = A^{p,q}(M, \mathbb{C}) \otimes E$. With respect to a local holomorphic frame $\Phi = (\Phi_j)$ of E, a differential form of type (p, q) with values in E is a linear combination

$$\alpha = \varphi^j \otimes \Phi_j, \tag{3.3}$$

where the coefficients φ^j are complex valued differential forms of type (p, q). The space of such differential forms is denoted $\mathcal{A}^{p,q}(M, E)$. We define the $\overline{\partial}$-operator on differential forms with values in E by

$$\overline{\partial}\alpha := (\overline{\partial}\varphi^j) \otimes \Phi_j. \tag{3.4}$$

Since the transition maps between holomorphic frames of E are holomorphic, it follows that $\overline{\partial}\alpha$ is well defined. By definition, $\overline{\partial}\alpha$ is of type $(p, q+1)$. We have $\overline{\partial}\overline{\partial} = 0$ and hence

$$\cdots \xrightarrow{\overline{\partial}} \mathcal{A}^{p,q-1}(M, E) \xrightarrow{\overline{\partial}} \mathcal{A}^{p,q}(M, E) \xrightarrow{\overline{\partial}} \mathcal{A}^{p,q+1}(M, E) \xrightarrow{\overline{\partial}} \cdots \tag{3.5}$$

is a cochain complex. The cohomology of this complex is denoted $H^{p,*}(M, E)$ and called *Dolbeault cohomology* of M with coefficients in E. The dimensions

$$h^{p,q}(M, E) := \dim_{\mathbb{C}} H^{p,q}(M, E) \tag{3.6}$$

are called *Hodge numbers* of M with respect to E.

3.7 Remark. Let $\Omega^p(E)$ be the sheaf of germs of holomorphic differential forms with values in E and of degree p (that is, of type $(p, 0)$). Let $\mathcal{A}^{p,q}(E)$ be the sheaf of germs of differential forms with values in E and of type (p, q). Then

$$0 \to \Omega^p(E) \hookrightarrow \mathcal{A}^{p,0}(E) \xrightarrow{\overline{\partial}} \mathcal{A}^{p,1}(E) \xrightarrow{\overline{\partial}} \mathcal{A}^{p,2}(E) \xrightarrow{\overline{\partial}} \cdots$$

is a fine resolution of $\Omega^p(E)$. Hence $H^q(M, \Omega^p(E))$ is isomorphic to $H^{p,q}(M, E)$.

We are going to use the notation and results from Chapter 1. We assume that M is endowed with a compatible metric in the sense of (2.33) and that E is endowed with a Hermitian metric. Then the splitting of forms into types as in (3.2) is perpendicular with respect to the induced Hermitian metric on $A^*(M, E)$.

Suppose that α and β are differential forms with values in E and of type (p, q) and $(p, q-1)$, respectively. By Corollary 2.36, $(\bar{*} \otimes h)\alpha$ is of type $(m-p, m-q)$. Therefore $((\bar{*} \otimes h)\alpha) \wedge_\varepsilon \beta$ is a complex valued differential form of type $(m, m-1)$. It follows that

$$\bar{\partial}\big(((\bar{*} \otimes h)\alpha) \wedge_\varepsilon \beta\big) = d\big(((\bar{*} \otimes h)\alpha) \wedge_\varepsilon \beta\big).$$

Since the real dimension of M is $2m$, we have $(\bar{*} \otimes h^*)(\bar{*} \otimes h) = (-1)^r$ on forms of degree $r = p + q$. Computing as in (1.46), we get

$$\begin{aligned} d\big(((\bar{*} \otimes h)\alpha) \wedge_\varepsilon \beta\big) &= \bar{\partial}\big(((\bar{*} \otimes h)\alpha) \wedge_\varepsilon \beta\big) \\ &= (-1)^{2m-r}\{-((\bar{*} \otimes h^*)\bar{\partial}(\bar{*} \otimes h)\alpha, \beta) + (\alpha, \bar{\partial}\beta)\} \operatorname{vol}. \end{aligned}$$

It follows that

$$\bar{\partial}^* = (\bar{*} \otimes h^*)\bar{\partial}(\bar{*} \otimes h). \tag{3.8}$$

Here the $\bar{\partial}$-operator on the right belongs to the dual bundle E^* of E. Let

$$\Delta_{\bar{\partial}} = \bar{\partial}\bar{\partial}^* + \bar{\partial}^*\bar{\partial} \tag{3.9}$$

be the Laplace operator associated to $\bar{\partial}$. We claim that

$$\Delta_{\bar{\partial}} (\bar{*} \otimes h) = (\bar{*} \otimes h)\, \Delta_{\bar{\partial}}, \tag{3.10}$$

where the Laplace operator on the left belongs to E^*. In fact,

$$\begin{aligned} \Delta_{\bar{\partial}} (\bar{*} \otimes h) &= \bar{\partial}(\bar{*} \otimes h)\bar{\partial}(\bar{*} \otimes h^*)(\bar{*} \otimes h) + (\bar{*} \otimes h)\bar{\partial}(\bar{*} \otimes h^*)\bar{\partial}(\bar{*} \otimes h) \\ &= (\bar{*} \otimes h)\bar{\partial}(\bar{*} \otimes h^*)\bar{\partial}(\bar{*} \otimes h) + (\bar{*} \otimes h)(\bar{*} \otimes h^*)\bar{\partial}(\bar{*} \otimes h)\bar{\partial} \\ &= (\bar{*} \otimes h)\, \Delta_{\bar{\partial}}. \end{aligned}$$

We denote by $\mathcal{H}^{p,q}(M, E)$ the space of $\Delta_{\bar{\partial}}$-harmonic forms of type (p, q) with coefficients in E. By (3.10), $\bar{*} \otimes h$ restricts to a conjugate linear isomorphism

$$\mathcal{H}^{p,q}(M, E) \to \mathcal{H}^{m-p,m-q}(M, E^*). \tag{3.11}$$

In the rest of this subsection, suppose that M is closed. Then by Hodge theory, the canonical projection $\mathcal{H}^{p,q}(M, E) \to H^{p,q}(M, E)$ is an isomorphism of vector spaces. From (3.11) we infer *Serre duality*, namely that $\bar{*} \otimes h$ induces a conjugate linear isomorphism

$$H^{p,q}(M, E) \to H^{m-p,m-q}(M, E^*). \tag{3.12}$$

In particular, we have $h^{p,q}(M, E) = h^{m-p,m-q}(M, E^*)$ for all p and q.

3.13 Remark. Equivalently we can say that $\bar{*} \otimes h$ induces a conjugate linear isomorphism $H^q(M, \Omega^p(E)) \cong H^{m-q}(M, \Omega^{m-p}(E^*))$, see Remark 3.7.

3.2 Chern Connection. Let D be a connection on E and σ a smooth section of E. Then we can split

$$D\sigma = D'\sigma + D''\sigma \tag{3.14}$$

with $D'\sigma \in \mathcal{A}^{1,0}(M,E)$ and $D''\sigma \in \mathcal{A}^{0,1}(M,E)$. Let $(\Phi_1,\dots,\Phi_k)$ be a holomorphic frame of E over an open subset $U \subset M$. Then

$$D\Phi_\mu = \theta^\nu_\mu \Phi_\nu, \tag{3.15}$$

where $\theta = (\theta^\nu_\mu)$ is the connection form and $\theta^\nu_\mu \Phi_\nu$ is shorthand for $\theta^\nu_\mu \otimes \Phi_\nu$.

3.16 Lemma. *$D''\sigma = 0$ for all local holomorphic sections σ of E iff $D'' = \bar\partial$. Then the connection form $\theta \in \mathcal{A}^{1,0}(U, \mathbb{C}^{k\times k})$.*

Proof. The assertion follows from comparing types in

$$D'\sigma + D''\sigma = D\sigma = d\varphi^\mu \Phi_\mu + \varphi^\mu D\Phi_\mu = ((\partial + \bar\partial)\varphi^\mu + \varphi^\nu \theta^\mu_\nu)\Phi_\mu,$$

where $\sigma = \varphi^\mu \Phi_\mu$ in terms of a local holomorphic frame (Φ_μ) of E. □

3.17 Lemma. *Let $h = (\cdot,\cdot)$ be a smooth Hermitian metric on E. If $D'' = \bar\partial$, then D is Hermitian iff*

$$(\sigma, D'_X\tau) = \partial(\sigma,\tau)(X) \quad \textit{or, equivalently,} \quad (D'_X\sigma, \tau) = \bar\partial(\sigma,\tau)(X)$$

for all vector fields X on M and local holomorphic sections σ, τ of E.

Proof. In the sense of 1-forms we write $(\sigma, D'\tau) = \partial(\sigma,\tau)$ for the equality in the lemma. If this equality holds, then

$$\begin{aligned} d(\sigma,\tau) &= \partial(\sigma,\tau) + \bar\partial(\sigma,\tau) \\ &= (D'\sigma,\tau) + (\sigma, D'\tau) \\ &= (D\sigma,\tau) + (\sigma, D\tau), \end{aligned}$$

where we use $D''\sigma = D''\tau = 0$. Since E has local holomorphic frames, we conclude that the equality in the lemma implies that D is Hermitian. The other direction is similar. □

3.18 Theorem. *Let $E \to M$ be a holomorphic vector bundle and $h = (\cdot,\cdot)$ be a smooth Hermitian metric on E. Then there is precisely one connection D on E such that* 1) *D is Hermitian and* 2) *$D'' = \bar\partial$.*

Proof. Let $(\Phi_1,\dots,\Phi_k)$ be a holomorphic frame of E over an open subset $U \subset M$ and $\theta = (\theta^\nu_\mu)$ be the corresponding connection form as in (3.15). Now 2) implies that θ is of type $(1,0)$. By Property 1),

$$\begin{aligned} dh_{\mu\nu} = dh(\Phi_\mu,\Phi_\nu) &= h(\theta^\lambda_\mu \Phi_\lambda, \Phi_\nu) + h(\Phi_\mu, \theta^\lambda_\nu \Phi_\lambda) \\ &= \bar\theta^\lambda_\mu h_{\lambda\nu} + h_{\mu\lambda}\theta^\lambda_\nu = \overline{h_{\nu\lambda}\theta^\lambda_\mu} + h_{\mu\lambda}\theta^\lambda_\nu. \end{aligned}$$

Now $dh_{\mu\nu} = \partial h_{\mu\nu} + \bar{\partial} h_{\mu\nu}$. Since θ is of type $(1,0)$, we conclude

$$\partial h_{\mu\nu} = h_{\mu\lambda}\theta^{\lambda}_{\nu} \quad \text{or} \quad \theta = h^{-1}\partial h, \tag{3.19}$$

by comparison of types. This shows that Properties 1) and 2) determine D uniquely. Existence follows from the fact that the local connection forms $\theta = h^{-1}\partial h$ above transform correctly under changes of frames. Another argument for the existence is that for local holomorphic sections σ and τ of E, $\partial(\sigma, \tau)$ is conjugate $\mathcal{O}$-linear in σ so that the equation in Lemma 3.17 leads to the determination of the yet undetermined D'. □

The unique connection D satisfying the properties in Theorem 3.18 will be called the *Chern connection*. It depends on the choice of a Hermitian metric on E.

3.20 Exercises. 1) Let $E \to M$ be a holomorphic vector bundle and h be a Hermitian metric on E. Let $f\colon N \to M$ be a holomorphic map. Then the Chern connection of the pull back Hermitian metric f^*h on the pull back $f^*E \to N$ is the pull back of the Chern connection on E with respect to h.

2) Let $E \to M$ be a holomorphic vector bundle and $E' \to M$ be a holomorphic vector subbundle of E. Let h be a Hermitian metric on E and h' be the restriction of h to E'. For a section σ of E write $\sigma = \sigma' + \sigma''$, where σ' is a section of E' and σ'' is perpendicular to E'. Let D be the Chern connection on E with respect to h. Then the Chern connection D' of E' with respect to h' is given by $D'\sigma = (D\sigma)'$.

Recall that this is the standard recipe of getting a Hermitian connection for a subbundle of a Hermitian bundle with a Hermitian connection.

In what follows, we use the wedge product as defined in (1.15) and the discussion in Example 1.16, applied to the trivial bundle $E = M \times \mathbb{C}^k$.

3.21 Proposition. *Let $E \to M$ be a holomorphic vector bundle with Hermitian metric h and corresponding Chern connection D. Let $(\Phi_1, \dots, \Phi_k)$ be a local holomorphic frame of E and θ and Θ be the corresponding connection and curvature form. Then:*

1) $\theta = h^{-1}\partial h$ *with* $h = (h_{\mu\nu})$*;*

2) $D' = \partial + \theta$ *and* $D'' = \bar{\partial}$*;*

3) θ *is of type* $(1,0)$ *and* $\partial\theta = -\theta \wedge_\mu \theta$*;*

4) $\Theta = \bar{\partial}\theta$ *and* Θ *is of type* $(1,1)$*;*

5) $\bar{\partial}\Theta = 0$ *and* $\partial\Theta = \Theta \wedge_\lambda \theta = \Theta \wedge_\mu \theta - \theta \wedge_\mu \Theta$.

In the case of a holomorphic line bundle $E \to M$, a local nowhere vanishing holomorphic section Φ, and $h = (\Phi, \Phi)$, Proposition 3.21 gives

$$\theta = \partial \ln h \quad \text{and} \quad \Theta = \bar{\partial}\partial \ln h. \tag{3.22}$$

We will use these formulas without further reference.

Proof of Proposition 3.21. The first assertion follows from the proof of Theorem 3.18, the second from Lemma 3.16. As for the third assertion, we have

$$\begin{aligned}\partial\theta &= \partial(h^{-1}\partial h) = -(h^{-1}\partial h h^{-1}) \wedge_\mu \partial h \\ &= -(h^{-1}\partial h) \wedge_\mu (h^{-1}\partial h) = -\theta \wedge_\mu \theta.\end{aligned}$$

In particular,

$$\Theta = \partial\theta + \bar\partial\theta + \theta \wedge_\mu \theta = \bar\partial\theta.$$

Hence $\bar\partial\Theta = 0$ and, by the Bianchi identity $d\Theta = \Theta \wedge_\lambda \theta$, we conclude that

$$\Theta \wedge_\lambda \theta = d\Theta = \partial\Theta + \bar\partial\Theta = \partial\Theta. \qquad \square$$

3.23 Proposition. *Let $E \to M$ be a holomorphic vector bundle with Hermitian metric h and Chern connection D. Let $p_0 \in M$ and z be holomorphic coordinates about p_0 with $z(p_0) = 0$. Then there is a holomorphic frame $(\Phi_1, \dots, \Phi_k)$ of E about p_0 such that*

1) $h(z) = 1 + O(|z|^2)$;

2) $\Theta(0) = \bar\partial\partial h(0)$.

Proof. Suppose 1) holds. Then

$$\Theta(0) = \bar\partial\theta(0) = \big((\bar\partial h^{-1}) \wedge_\mu \partial h + h^{-1}\bar\partial\partial h\big)(0) = \bar\partial\partial h(0),$$

hence 1) implies 2). To show 1), we choose a holomorphic frame $(\Phi_1, \dots, \Phi_k)$ about p_0 such that $h_{\mu\nu}(0) = \delta_{\mu\nu}$. We define a new holomorphic frame

$$\tilde\Phi_\mu = \Phi_\mu + z^i a^\nu_{\mu i}\Phi_\nu$$

with $a^\nu_{\mu i} = -(\partial h_{\nu\mu}/\partial z^i)(0)$. It is easy to check that $\tilde h = 1 + O(|z|^2)$. $\square$

3.24 Example (Tautological bundle). Let $M = G_{r,n} := G_r\mathbb{C}^n$ and $E \to M$ be the tautological bundle as in Example 2.2.5. The standard Hermitian inner product on $\mathbb{C}^n$ induces a Hermitian metric on E.

We identify sections of E over an open subset U of M with maps $\sigma\colon U \to \mathbb{C}^n$ such that $\sigma(p) \in p$ for all $p \in U$. Then a holomorphic frame over an open subset $U \subset M$ is given by holomorphic mappings $\Phi_\mu\colon U \to \mathbb{C}^n$, $1 \le \mu \le r$, such that $(\Phi_1(p), \dots, \Phi_r(p))$ is a basis of p, for all $p \in U$. For such a frame,

$$h_{\mu\nu} = (\Phi_\mu, \Phi_\nu) = \sum_\lambda \bar\Phi^\lambda_\mu \Phi^\lambda_\nu = (\bar\Phi^t\Phi)^\mu_\nu \quad \text{with } \Phi = (\Phi^\nu_\mu).$$

If $\sigma\colon U \to \mathbb{C}^m$ is a section of E and X is a vector field of M over U, then the derivative $d\sigma(X)$ of σ in the direction of X need not be a section of E; that is, $d\sigma_p(X_p)$ need not be an element of p anymore. We obtain the Chern connection D of E by setting

$$(D_X\sigma)(p) = \pi_p(d\sigma_p(X_p)),$$

where $\pi_p\colon \mathbb{C}^n \to \mathbb{C}^n$ is the orthogonal projection of $\mathbb{C}^n$ onto p, compare Exercise 3.20.2 (where the ambient bundle is $M \times \mathbb{C}^n$ in our case).

Consider the plane p_0 spanned by the first r unit vectors, in homogeneous coordinates in $\mathbb{C}^{n\times r}$ written as

$$p_0 = \begin{bmatrix} 1 \\ 0 \end{bmatrix} \in M,$$

where $1 = 1_r$ is the $r \times r$ unit matrix. Then we get a holomorphic parameterization (the inverse of holomorphic coordinates) of M about p_0 by

$$\mathbb{C}^{(n-r)\times r} \ni z \mapsto \begin{bmatrix} 1 \\ z \end{bmatrix} \in M.$$

Moreover, the r columns of the matrix $\binom{1}{z}$ are a holomorphic frame of E. For this frame we have

$$h_{\mu\nu} = \delta_{\mu\nu} + \sum_\lambda \bar{z}^\lambda_\mu z^\lambda_\nu = \delta_{\mu\nu} + (\bar{z}^t z)^\mu_\nu \,.$$

In particular,

$$h = 1 + O(|z|^2).$$

In $z = 0$, that is, in p_0, we have

$$\Theta(0) = \bar\partial\partial h(0) = \bar\partial\bar{z}^t \wedge \partial z = d\bar{z}^t \wedge dz.$$

The canonical action of the group $\mathrm{U}(n)$ on M is transitive and extends canonically to an action on E which leaves h invariant. Each element from $\mathrm{U}(n)$ acts biholomorphically on M and E. In particular, choosing p_0 as above means no restriction of generality.

3.3 Some Formulas. For the following, we refer again to notation and results introduced in Chapter 1. Let $E \to M$ be a holomorphic vector bundle over M. Let h be a Hermitian metric on E and D be the corresponding Chern connection. Then we have the associated exterior differential

$$d^D\colon \mathcal{A}^r(M, E) \to \mathcal{A}^{r+1}(M, E), \tag{3.25}$$

see (1.18) and (1.21). Let (Φ_μ) be a local holomorphic frame of E and

$$\alpha = \varphi^\mu \Phi_\mu \tag{3.26}$$

be a differential form with values in E of type (p, q) with $p + q = r$. By the characteristic property of the Chern connection, we have

$$d^D\alpha = d\varphi^\mu \otimes \Phi_\mu + (-1)^r \varphi^\mu \wedge D\Phi_\mu = \partial^D\alpha + \bar\partial\alpha, \tag{3.27}$$

where

$$\partial^D\alpha := \partial\varphi^\mu \otimes \Phi_\mu + (-1)^r \varphi^\mu \wedge D\Phi_\mu. \tag{3.28}$$

Note that $\partial^D\alpha$ and $\bar\partial\alpha$ are of type $(p + 1, q)$ and $(p, q + 1)$, respectively.

3.29 Proposition. *Let E be a holomorphic vector bundle over M with Hermitian metric and associated Chern connection D. Then, for any differential form α with values in E,*

$$(\partial^D)^2\alpha = (\bar\partial)^2\alpha = 0 \quad \text{and} \quad (\partial^D\bar\partial + \bar\partial\partial^D)\alpha = (d^D)^2\alpha = R^D \wedge_\varepsilon \alpha,$$

where $\varepsilon\colon \operatorname{End}(E) \otimes E \to E$ is the evaluation map.

Proof. If α is a differential form with values in E of type (p,q), then $d^Dd^D\alpha = R^D \wedge_\varepsilon \alpha$, see Proposition 1.19. By Proposition 3.21, R^D is of type $(1,1)$, hence $d^Dd^D\alpha$ is of type $(p+1,q+1)$. By definition,

$$d^Dd^D\alpha = \partial^D\partial^D\alpha + \bar\partial\bar\partial\alpha + (\partial^D\bar\partial + \bar\partial\partial^D)\alpha.$$

The first two forms on the right hand side vanish since they have type $(p+2,q)$ and $(p,q+2)$, respectively. □

We assume now that, in addition, M is endowed with a compatible Riemannian metric as in (2.33). Let $(X_1, Y_1, \dots, X_m, Y_m)$ be a local orthonormal frame of M with $JX_j = Y_j$. Set

$$Z_j = \frac{1}{2}(X_j - iY_j) \quad \text{and} \quad \bar Z_j = \frac{1}{2}(X_j + iY_j), \tag{3.30}$$

and let $Z_1^*, \dots, Z_n^*, \bar Z_1^*, \dots, \bar Z_n^*$ be the corresponding dual frame of $T^*_{\mathbb{C}}M$. With our conventions, we have

$$Z_j^\flat := \langle Z_j, \cdot\rangle = \frac{1}{2}\bar Z_j^* \quad \text{and} \quad \bar Z_j^\flat := \langle \bar Z_j, \cdot\rangle = \frac{1}{2}Z_j^*. \tag{3.31}$$

Since $X_j = Z_j + \bar Z_j$ and $Y_j = i(Z_j - \bar Z_j)$, we get

$$X_j^* = \frac{1}{2}(Z_j^* + \bar Z_j^*) \quad \text{and} \quad Y_j^* = -\frac{i}{2}(Z_j^* - \bar Z_j^*)$$

for the dual frame $(X_1^*, Y_1^*, \dots, X_m^*, Y_m^*)$ of $(X_1, Y_1, \dots, X_m, Y_m)$. Therefore,

$$\begin{aligned} d^D &= \frac{1}{2}\sum \{(Z_j^* + \bar Z_j^*) \wedge \hat D_{Z_j + \bar Z_j} + (Z_j^* - \bar Z_j^*) \wedge \hat D_{Z_j - \bar Z_j}\} \\ &= \sum \{Z_j^* \wedge \hat D_{Z_j} + \bar Z_j^* \wedge \hat D_{\bar Z_j}\} = \partial^D + \bar\partial, \end{aligned} \tag{3.32}$$

meaning that the sum of the first terms is equal to ∂^D and the sum of the second to $\bar\partial$. To compute the adjoint operator $(d^D)^*$ of d^D with respect to the induced Hermitian metric as in (1.25), we note first that the relations in Exercise 1.26 also hold for complex tangent vectors. Since $Z_j^* = 2\bar Z_j^\flat$ and $\bar Z_j^* = 2Z_j^\flat$, we get

$$(d^D)^* = -2\sum \{Z_j \lrcorner \hat D_{\bar Z_j} + \bar Z_j \lrcorner \hat D_{Z_j}\} = (\partial^D)^* + \bar\partial^*, \tag{3.33}$$

where we use the first relation from Exercise 1.26 for the first equality and comparison of types for the second. The Laplace operators associated to $\bar\partial$ and ∂^D are defined by

$$\Delta_{\bar\partial} = \bar\partial\bar\partial^* + \bar\partial^*\bar\partial \quad \text{and} \quad \Delta_{\partial^D} = \partial^D(\partial^D)^* + (\partial^D)^*\partial^D, \tag{3.34}$$

respectively. Both preserve the type of forms. Using the second relation from Exercise 1.26, a straightforward computation gives the following Weitzenböck formulas,

$$\begin{aligned}\Delta_{\bar\partial}\alpha &= -2\sum_j \hat D^2\alpha(Z_j, \bar Z_j) + 2\sum_{j,k} \bar Z_j^* \wedge (\bar Z_k \llcorner (\hat R^D(Z_k, \bar Z_j)\alpha)) \\ &= -2\sum_j \hat D^2\alpha(\bar Z_j, Z_j) + 2\sum_{j,k} \bar Z_j \llcorner (\bar Z_k^* \wedge (\hat R^D(\bar Z_k, Z_j)\alpha)),\end{aligned} \tag{3.35}$$

$$\begin{aligned}\Delta_{\partial^D}\alpha &= -2\sum_j \hat D^2\alpha(Z_j, \bar Z_j) + 2\sum_{j,k} Z_j \llcorner (Z_k^* \wedge (\hat R^D(Z_k, \bar Z_j)\alpha)) \\ &= -2\sum_j \hat D^2\alpha(\bar Z_j, Z_j) + 2\sum_{j,k} Z_j^* \wedge (Z_k \llcorner (\hat R^D(\bar Z_k, Z_j)\alpha)),\end{aligned} \tag{3.36}$$

where the frame $(Z_1, \bar Z_1, \dots, Z_m, \bar Z_m)$ is as in (3.30) and where $\hat R^D$ denotes the curvature tensor of $\hat D$ on $A^*(M, E)$.

3.37 Exercises. 1) Show that, for a differential form $\varphi \otimes \sigma$ of degree r,

$$\begin{aligned}\partial^D(\varphi \otimes \sigma) &= \partial\varphi \otimes \sigma + (-1)^r \varphi \wedge D'\sigma, \\ \bar\partial(\varphi \otimes \sigma) &= \bar\partial\varphi \otimes \sigma + (-1)^r \varphi \wedge D''\sigma.\end{aligned}$$

2) Prove Formulas 3.35 and 3.36 and show that

$$\begin{aligned}4\sum \hat D^2\alpha(Z_j, \bar Z_j) &= \operatorname{tr} \hat D^2\alpha + i\sum \hat R^D(X_j, JX_j)\alpha, \\ 4\sum \hat D^2\alpha(\bar Z_j, Z_j) &= \operatorname{tr} \hat D^2\alpha + i\sum \hat R^D(JX_j, X_j)\alpha.\end{aligned}$$

3.4 Holomorphic Line Bundles. The results in this subsection will be mainly used in Section 9. We continue to assume that M is a complex manifold. Let $E \to M$ be a complex line bundle. Let (U_α) be an open covering of M such that, for each α, there is a nowhere vanishing smooth section $\Phi_\alpha \colon U_\alpha \to E$. If σ is a smooth section of E, then we can write $\sigma = \sigma_\alpha \Phi_\alpha$ over U_α, where $\sigma_\alpha \colon U_\alpha \to \mathbb{C}$ is a smooth function, the *principal part* of σ with respect to Φ_α.

Over intersections $U_\alpha \cap U_\beta$, we have $\Phi_\beta = t_{\alpha\beta}\Phi_\alpha$, where the transition functions $t_{\alpha\beta} \colon U_\alpha \cap U_\beta \to \mathbb{C}^* = \mathbb{C} \setminus \{0\}$ are smooth. For the principal parts of a section σ as above we get

$$\sigma_\alpha = t_{\alpha\beta}\sigma_\beta. \tag{3.38}$$

The family $(t_{\alpha\beta})$ of transition functions satisfies

$$t_{\alpha\alpha} = t_{\alpha\beta}t_{\beta\alpha} = t_{\alpha\beta}t_{\beta\gamma}t_{\gamma\alpha} = 1. \tag{3.39}$$

We say that with respect to the given covering, $(t_{\alpha\beta})$ is a 1-cocycle of smooth functions with values in $\mathbb{C}^*$.

Suppose E' is another complex line bundle over M, and suppose there is an isomorphism, $F\colon E \to E'$. After passing to a common refinement if necessary, suppose that we are given, for each α, a nowhere vanishing smooth section $\Phi'_\alpha\colon U_\alpha \to E'$. As above, we have smooth transition functions $(t'_{\alpha\beta})$ with $\Phi'_\beta = t'_{\alpha\beta}\Phi'_\alpha$. Over each U_α, we also have $F(\Phi_\alpha) = s_\alpha \Phi'_\alpha$, where $s_\alpha\colon U_\alpha \to \mathbb{C}^*$ is smooth. Comparing coefficients we get

$$t'_{\alpha\beta} = s_\alpha t_{\alpha\beta} s_\beta^{-1}. \tag{3.40}$$

By definition, this means that the two cocycles $(t_{\alpha\beta})$ and $(t'_{\alpha\beta})$ are cohomologous.

Vice versa, suppose (U_α) is an open covering of M and $(t_{\alpha\beta})$ a $\mathbb{C}^*$-valued 1-cocycle of smooth functions with respect to (U_α). On the set of triples (α, p, v) with $p \in U_\alpha$ and $v \in \mathbb{C}$ set $(\alpha, p, v) \sim (\beta, q, w)$ iff $p = q$ and $v = t_{\alpha\beta} w$. By (3.39), $\sim$ is an equivalence relation. The set E of equivalence classes $[\alpha, p, v]$ admits a natural projection to M, $\pi([\alpha, p, v]) = p$. The fibers $\pi^{-1}(p)$, $p \in M$, are complex lines. As in the case of the universal bundles over complex Grassmannians, it follows that there is a unique topology on E such that $\pi\colon E \to M$ is a smooth complex line bundle with nowhere vanishing smooth sections

$$\Phi_\alpha\colon U_\alpha \to E, \quad \Phi_\alpha(p) = [\alpha, p, 1]. \tag{3.41}$$

If another cocycle $t'_{\alpha\beta}$ is cohomologous to the given one, $t'_{\alpha\beta} = s_\alpha t_{\alpha\beta} s_\beta^{-1}$ for a family of smooth functions $s_\alpha\colon U_\alpha \to \mathbb{C}^*$, then $F(\Phi_\alpha) := s_\alpha \Phi'_\alpha$ gives rise to an isomorphism $F\colon E \to E'$.

In the language of sheaves, we have established that the space of isomorphism classes of smooth complex line bundles over M is naturally isomorphic to $H^1(M, \mathcal{E}^*)$, where $\mathcal{E}^*$ is the sheaf over M of germs of smooth functions with values in $\mathbb{C}^*$.

Replacing the word 'smooth' in the above discussion by the word 'holomorphic', we obtain the corresponding results for holomorphic line bundles. In particular, the space of isomorphism classes of holomorphic line bundles over M is naturally isomorphic to the *Picard group* $\mathrm{Pic}(M) := H^1(M, \mathcal{O}^*)$, where $\mathcal{O}^*$ is the sheaf of germs of holomorphic functions with values in $\mathbb{C}^*$ and where the group law in $\mathrm{Pic}(M)$ corresponds to the tensor product of line bundles. Note that the tensor product $L \otimes L^*$ of a holomorphic line bundle L over M with its dual bundle L^* is holomorphically isomorphic to the trivial bundle so that the dual bundle corresponds to the inverse in $\mathrm{Pic}(M)$.

There is an important connection between complex hypersurfaces and holomorphic line bundles. Suppose H is a complex hypersurface. Let (U_α) be an open covering of M with defining holomorphic functions $f_\alpha\colon U_\alpha \to \mathbb{C}$ for H,

$$H \cap U_\alpha = \{p \in U_\alpha \mid f_\alpha(p) = 0\}, \tag{3.42}$$

such that $df_\alpha(p) \neq 0$ for all $p \in H \cap U_\alpha$. Then we have

$$f_\alpha = t_{\alpha\beta} f_\beta \tag{3.43}$$

over intersections $U_\alpha \cap U_\beta$, where the functions $t_{\alpha\beta}\colon U_\alpha \cap U_\beta \to \mathbb{C}^*$ are holomorphic. The family $(t_{\alpha\beta})$ is a 1-cocycle of holomorphic functions and, therefore, defines a holomorphic line bundle E as above such that the sections Φ_α from (3.41) are holomorphic and such that the section f of E with $f|U_\alpha = f_\alpha \Phi_\alpha$ is well-defined and holomorphic. In this way we associate to H a holomorphic line bundle E together with a holomorphic section f such that $H = \{p \in M \mid f(p) = 0\}$.

3.44 Remark. More generally, there are one-to-one correspondences between so-called effective divisors and pairs consisting of a line bundle together with a holomorphic section (up to isomorphism) respectively divisors and pairs of a line bundle together with a meromorphic section (up to isomorphism), see for example Chapitre V in [Wei], Section 1.1 in [GH], or Section 2.3, Proposition 4.4.13, and Corollary 5.3.7 in [Hu].

3.45 Examples. We consider some line bundles over $\mathbb{C}P^m$. We will use the open covering of $\mathbb{C}P^m$ by the sets $U_i = \{[z] \in \mathbb{C}P^m \mid z_i \neq 0\}$, $0 \le i \le m$.

1) The tautological line bundle $U = U_{1,m} \to \mathbb{C}P^m$, compare also Examples 2.2.5 and 3.24: Over U_i,

$$\Phi_i = \left(\frac{z_0}{z_i}, \dots, \frac{z_i}{z_i}, \dots, \frac{z_m}{z_i}\right)$$

is a nowhere vanishing holomorphic section of U over U_i. Over intersections $U_i \cap U_j$, we have $z_i \Phi_i = z_j \Phi_j$.

Over U_i, a section σ of U is of the form $\sigma = \sigma_i \Phi_i$, where σ_i is a function called the principal part of σ with respect to Φ_i. Over intersections $U_i \cap U_j$, we have

$$\sigma_i = \frac{z_i}{z_j}\sigma_j.$$

2) The canonical line bundle $K = A^{m,0}(\mathbb{C}P^m, \mathbb{C}) \to \mathbb{C}P^m$: Over U_i as above, let $w_{ij} = z_j/z_i$ and set

$$\omega_i = (-1)^i dw_{i0} \wedge \cdots \wedge \widehat{dw_{ii}} \wedge \cdots \wedge dw_{im},$$

where the hat indicates that the corresponding term is to be deleted. Then ω_i is a nowhere vanishing holomorphic section of K over U_i. Over $U_i \cap U_j$,

$$w_{ik} = \frac{z_k}{z_j} \cdot \frac{z_j}{z_i} = \frac{z_j}{z_i} \cdot w_{jk} \quad \text{and} \quad d\frac{z_j}{z_i} = -\frac{z_j^2}{z_i^2} d\frac{z_i}{z_j}.$$

Therefore

$$z_i^{m+1}\omega_i = z_j^{m+1}\omega_j,$$

and hence K is isomorphic to U^{m+1}, the $(m+1)$-fold tensor product $U \otimes \cdots \otimes U$.

3) A hyperplane in $\mathbb{C}P^m$ is determined by an m-dimensional vector subspace $V \subset \mathbb{C}^{m+1}$. We choose the hyperplane $H = \{[z] \in \mathbb{C}P^m \mid z_0 = 0\}$. Over U_i as above, H is defined by the equation

$$f_i([z]) := \frac{z_0}{z_i} = 0.$$

Over $U_i \cap U_j$,

$$f_i = \frac{z_j}{z_i} f_j.$$

For the holomorphic line bundle over $\mathbb{C}P^m$ associated to H, also denoted by H, the functions f_i serve as principal parts of a holomorphic section f of H which vanishes along the given hyperplane. It follows that in our case, H is inverse to U, that is, $H \otimes U$ is the trivial bundle.

4) Let M be a complex manifold of complex dimension m and $\tilde{M}$ be the blow up of M at a point $p \in M$ as in Subsection 2.4. We use the notation introduced there and describe the holomorphic line bundle $L \to \tilde{M}$ determined by the hypersurface $S = \pi^{-1}(p)$. To that end, we consider the open covering of $\tilde{M}$ consisting of the subsets

$$W_0 = \tilde{M} \setminus S, \quad W_i = \{(z, [w]) \in V \mid w_i \neq 0\},\ 1 \leq i \leq m,$$

of $\tilde{M}$ and corresponding defining holomorphic functions $f_i \colon W_i \to \mathbb{C}$ of S,

$$f_0 = 1 \quad \text{and} \quad f_i(z, [w]) = z_i,\ 1 \leq i \leq m.$$

These functions transform under the rules

$$z_j f_0 = f_j \quad \text{and} \quad z_j f_i = z_i f_j.$$

The functions f_i define a global holomorphic section f of L with S as its set of zeros. We have $f = f_i \Phi_i$, where Φ_i is a nowhere vanishing holomorphic section of L over W_i as in (3.41).

Let M be a complex manifold of complex dimension m and $\tilde{M}$ be the blow up of M at a point $p \in M$ as in Subsection 2.4. Let $L \to \tilde{M}$ be the holomorphic line bundle as in Example 3.45.4. In the notation of Subsection 2.4, consider the map $\sigma \colon V \to \mathbb{C}P^{m-1}$, $\sigma(z, [w]) = [w]$. Let $H \to \mathbb{C}P^{m-1}$ be the hyperplane bundle, see Example 3.45.3.

3.46 Lemma. *Let L^* be the dual of L. Then*

$$L^* | V \cong \sigma^* H.$$

Proof. Realize H as the line bundle over $\mathbb{C}P^{m-1}$ associated to the hyperplane $\{w_1 = 0\}$. Then the defining equations of H over W_i are $g_i = w_1/w_i = 0$. The functions g_i transform according to the rule $w_i g_i = w_j g_j$, or, equivalently, $z_i g_i = z_j g_j$. Hence $\sigma^* H$ is inverse to $L|V$. □

3.47 Lemma. *The canonical bundles $K = A^{m,0}(M, \mathbb{C})$ of M and $\tilde{K} = A^{m,0}(\tilde{M}, \mathbb{C})$ of $\tilde{M}$ are related by*

$$\tilde{K} \cong \pi^* K \otimes L^{m-1}.$$

Proof. For W_i, $i \geq 1$, as above, let $u_j = w_j/w_i = z_j/z_i$. Then

$$(u_1, \ldots, u_{i-1}, z_i, u_{i+1}, \ldots, u_m)$$

are holomorphic coordinates on W_i. Hence

$$\Psi_i = du_1 \wedge \cdots \wedge du_{i-1} \wedge dz_i \wedge du_{i+1} \wedge \cdots \wedge du_m$$

is a nowhere vanishing section of $\tilde{K}$ over W_i. Since $dz_j = u_j dz_i + z_i du_j$,

$$\Psi_i = z_i^{1-m} \, dz_1 \wedge \cdots \wedge dz_m.$$

Hence we have

$$z_i^{m-1} \Psi_i = z_j^{m-1} \Psi_j$$

over $W_i \cap W_j$. Similarly, $\Psi_0 = dz_1 \wedge \cdots \wedge dz_m$ vanishes nowhere over $W_0 \cap W_j$ and

$$\Psi_0 = z_j^{m-1} \Psi_j.$$

Over V, we view $dz_1 \wedge \cdots \wedge dz_m$ also as a nowhere vanishing section of $\pi^* K$.

Now L is trivial over $W_0 = \tilde{M} \setminus S$ with nowhere vanishing section Φ_0 as above. Moreover, the differential of π induces an isomorphism, also denoted π^*, between K restricted to $M \setminus \{p\}$ and $\tilde{K}$ restricted to W_0. Hence over W_0, we obtain an isomorphism as desired by sending $\pi^*\sigma$ to $(\sigma \circ \pi) \otimes \Phi_0^{m-1}$. Similarly, we obtain an isomorphism over W_i, $i \geq 1$, by sending Ψ_i to $(dz_1 \wedge \cdots \wedge dz_m) \otimes \Phi_i^{m-1}$. By the choice of sections Ψ_i, these isomorphisms agree on the intersections $W_i \cap W_j$, hence define an isomorphism over $\tilde{M}$. □

4 Kähler Manifolds

Let M be a complex manifold with complex structure J and compatible Riemannian metric $g = \langle \cdot, \cdot \rangle$ as in (2.33). The alternating 2-form

$$\omega(X, Y) := g(JX, Y) \tag{4.1}$$

is called the associated *Kähler form*. We can retrieve g from ω,

$$g(X, Y) = \omega(X, JY). \tag{4.2}$$

We say that g is a *Kähler metric* and that M (together with g) is a *Kähler manifold* if ω is closed[9].

4.3 Remark. View TM together with J as a complex vector bundle over M, and let h be a Hermitian metric on TM. Then $g = \operatorname{Re} h$ is a compatible Riemannian metric on M and $\operatorname{Im} h$ is the associated Kähler form:

$$g(JX, Y) = \operatorname{Re} h(JX, Y) = \operatorname{Re} h(iX, Y) = \operatorname{Re}(-ih(X, Y)) = \operatorname{Im} h(X, Y).$$

Vice versa, if g is a compatible Riemannian metric on M and ω is the associated Kähler form, then $h = g + i\omega$ is a Hermitian metric on TM.

In terms of holomorphic coordinates z on M and the frames introduced in (2.13), we have

$$g_{jk} = \langle Z_j, Z_k \rangle = 0, \quad g_{\bar{j}\bar{k}} = \langle \bar{Z}_j, \bar{Z}_k \rangle = 0 \tag{4.4}$$

and

$$g_{j\bar{k}} = \langle Z_j, \bar{Z}_k \rangle = \langle \bar{Z}_k, Z_j \rangle = g_{\bar{k}j}. \tag{4.5}$$

We also have

$$g_{\bar{j}k} = \bar{g}_{j\bar{k}}. \tag{4.6}$$

With

$$dz^j \odot d\bar{z}^k := dz^j \otimes d\bar{z}^k + d\bar{z}^k \otimes dz^j \tag{4.7}$$

we finally get

$$g = g_{j\bar{k}} dz^j \odot d\bar{z}^k, \quad \omega = i g_{j\bar{k}} dz^j \wedge d\bar{z}^k. \tag{4.8}$$

We call the matrix $(g_{j\bar{k}})$ the *fundamental matrix* of g (with respect to the given coordinates).

4.9 Remark. In terms of holomorphic coordinates z as above and the usual notation, we have

$$\begin{aligned} g_{j\bar{k}} &= \frac{1}{4}\langle X_j - iY_j, X_k + iY_k \rangle \\ &= \frac{1}{2}\big\{\langle X_j, X_k \rangle - i\omega(X_j, X_k)\big\} = \frac{1}{2}\bar{h}_{jk}, \end{aligned}$$

[9]Kähler's original article [Kä1] contains much of what we say in this chapter. Compare also Bourguignon's essay in [Kä2] or his review of Kähler's article in [Bo2].

where the h_{jk} are the coefficients of the fundamental matrix of h with respect to the local complex frame $(X_1, \dots, X_m)$ of TM.

4.10 Examples. 1) $M = \mathbb{C}^m$ with the Euclidean metric g. Then g is a Kähler metric with

$$g = \frac{1}{2}\sum dz^j \odot d\bar{z}^j, \quad \omega = \frac{i}{2}\sum dz^j \wedge d\bar{z}^j. \tag{4.11}$$

For any lattice $\Gamma \subset \mathbb{C}^m$, the induced metric on the complex torus $\Gamma\backslash\mathbb{C}^m$ is a Kähler metric.

2) Products of Kähler manifolds (endowed with the product complex structure and the product metric) are Kähler manifolds.

3) A complex submanifold N of a Kähler manifold M is Kählerian with respect to the induced Riemannian metric since the Kähler form on N is the restriction of the Kähler form of M. More generally, if $N \to M$ is a holomorphic immersion, then N is Kählerian with respect to the induced Riemannian metric.

4) When endowed with the induced complex structure and Riemannian metric, covering spaces of Kähler manifolds and quotients of Kähler manifolds by properly discontinuous and free group actions by holomorphic and isometric transformations are again Kähler manifolds.

5) The following construction yields the *Fubini–Study* metric in the case of complex projective spaces. Let $G_{r,n}$ be the Grassmannian of r-planes in $\mathbb{C}^n$, compare Examples 2.2.4, B.42, and B.83.1. Let $M^* \subset \mathbb{C}^{n\times r}$ be the open subset of matrices of rank r and $\pi \colon M^* \to G_{r,n}$ be the canonical projection, a holomorphic principal bundle with structure group $\mathrm{Gl}(r, \mathbb{C})$ acting on M^* on the right by matrix multiplication. Let Z be a holomorphic section of π over an open subset $U \subset G_{r,n}$. Define a closed form ω of type $(1, 1)$ on U by

$$\omega = i\partial\bar{\partial} \ln \det(\bar{Z}^t Z). \tag{4.12}$$

We show first that ω does not depend on the choice of Z. In fact, any other choice is of the form ZF, where $F \colon U \to \mathrm{Gl}(r, \mathbb{C})$ is holomorphic, and then

$$\partial\bar{\partial} \ln \det(\bar{F}^t \bar{Z}^t Z F) = \partial\bar{\partial} \ln \det(\bar{Z}^t Z)$$

since $\ln \det F$ is holomorphic and $\ln \det \bar{F}^t$ antiholomorphic. This shows that ω is well defined independently of Z, and hence ω is a smooth form of type $(1, 1)$ on all of $G_{r,n}$. The defining formula for ω shows that ω is closed.

In the case of complex projective space, that is, in the case $r = 1$, a straightforward computation gives an explicit formula for ω (and the Fubini–Study metric). Let $U = \{[z^0, z] \in \mathbb{C}P^{n-1} \mid z^0 \neq 0\}$. Then $w = (1, z/z^0)$ is a section of π over U and hence

$$\begin{aligned}\omega &= i\partial\bar{\partial} \ln \det(1 + |w|^2) \\ &= \frac{i}{(1+|w|^2)^2}\Big((1+|w|^2)\sum_j dw^j \wedge d\bar{w}^j - \sum_j \bar{w}^j dw^j \wedge \sum_j w^j d\bar{w}^j\Big).\end{aligned} \tag{4.13}$$

The case $n = 2$ of complex dimension 1 and the comparison with Formula 4.15 are instructive. In the case $1 < r < n - 1$, explicit formulas are more complicated.

The r-plane spanned by $Z \in M^*$ is denoted $[Z]$, that is, $[Z] = \pi(Z)$. We recall the holomorphic action of $\mathrm{Gl}(n, \mathbb{C})$ on $G_{r,n}$, see Example 2.2.4. For A in $\mathrm{Gl}(n, \mathbb{C})$, we set $\lambda_A([Z]) = [AZ]$ and show next that $\lambda_A^* \omega = \omega$ for any A in $\mathrm{U}(n)$. To that end we let Z be a holomorphic section of π over the open subset $U \subset G_{r,n}$. Then $A^{-1}Z \circ \lambda_A$ is a section of π over $\lambda_A^{-1}(U)$ and

$$\begin{aligned} \lambda_A^* \omega &= i \lambda_A^* (\partial \bar\partial \ln \det(\bar Z^t Z)) \\ &= i \partial \bar\partial (\lambda_A^* (\ln \det(\bar Z^t Z))) \\ &= i \partial \bar\partial (\ln \det((\bar Z^t A)(A^{-1} Z)) \circ \lambda_A) = \omega \end{aligned}$$

since λ_A is holomorphic and A is unitary.

We want to show now that $g(X, Y) = \omega(X, JY)$ is a Kähler metric on $G_{r,n}$. By our discussion above it remains to show that g is positive definite. Since the action of $\mathrm{U}(n)$ is transitive on $G_{r,n}$ and g is invariant under this action, it suffices to show this in the point p_0, the plane spanned by the first r unit vectors in $\mathbb{C}^n$. About p_0 we have holomorphic coordinates $\left[\begin{smallmatrix} 1 \\ w \end{smallmatrix}\right] \mapsto w \in \mathbb{C}^{(n-r)\times r}$, where 1 stands for the $r \times r$ unit matrix. Hence

$$\begin{aligned} \omega(0) &= i \partial \bar\partial \ln \det(1 + \bar w^t w)|_{w=0} \\ &= i \partial \operatorname{tr}(((\bar\partial \bar w^t) w)(1 + \bar w^t w)^{-1})|_{w=0} \\ &= -i \operatorname{tr} \bar\partial \bar w^t \wedge \partial w \\ &= -i \operatorname{tr} d\bar w^t \wedge dw = i \sum_{j,k} dw_j^k \wedge d\bar w_j^k. \end{aligned}$$

It follows that

$$g(0) = \sum_{j,k} dw_j^k \odot d\bar w_j^k, \tag{4.14}$$

and hence g is positive definite. In conclusion, g is a Kähler metric on $G_{r,n}$ with Kähler form ω and the natural action of $\mathrm{U}(n)$ on $G_{r,n}$ preserves g and ω.

6) This example is related to the previous one, compare again Examples B.42 and B.83.1. Let $G_{r,n}^- \subset G_{r,n}$ be the open subset of r-planes on which the Hermitian form $Q_{r,n-r}$ as in (B.49) is negative definite. Write $Z \in \mathbb{C}^{n\times r}$ as

$$Z = \begin{pmatrix} Z_0 \\ Z_1 \end{pmatrix} \quad \text{with } Z_0 \in \mathbb{C}^{r\times r} \text{ and } Z_1 \in \mathbb{C}^{(n-r)\times r}.$$

The columns of Z span an r-plane, $[Z]$, in $G_{r,n}^-$ iff

$$\bar Z^t Q_{r,n-r} Z = -\bar Z_0^t Z_0 + \bar Z_1^t Z_1$$

is negative definite, that is, iff $\bar Z_0^t Z_0 > \bar Z_1^t Z_1$ in the sense of Hermitian matrices. Then Z_0 is invertible. Consider the open subset

$$M^* = \{Z \in \mathbb{C}^{n\times r} \mid \bar Z_0^t Z_0 > \bar Z_1^t Z_1\} \subset \mathbb{C}^{n\times r}$$

and the canonical projection $\pi : M^* \to G^-_{r,n}$, a holomorphic principal bundle with structure group $\mathrm{Gl}(r, \mathbb{C})$ acting on M^* on the right by matrix multiplication. Define a closed form ω of type (1, 1) on $G^-_{r,n}$ by

$$\omega = -i\partial\bar{\partial} \ln \det(\bar{Z}_0^t Z_0 - \bar{Z}_1^t Z_1),$$

where Z is a local holomorphic section of π. As in the previous example we see that ω is well defined and positive definite.

Let $\mathrm{U}(r, n-r)$ be the group of linear transformations of $\mathbb{C}^n$ preserving $Q_{r,n-r}$. Then the natural action of $\mathrm{U}(r, n-r)$ on $G_{r,n}$ leaves $G^-_{r,n}$ invariant and is transitive on $G^-_{r,n}$. The stabilizer of the plane $p_0 \in G^-_{r,n}$ spanned by the first r unit vectors is $\mathrm{U}(r) \times \mathrm{U}(n-r)$. As in the previous example we see that ω is invariant under the action of $\mathrm{U}(r, n-r)$ and that $g(X, Y) = \omega(X, JY)$ is a Kähler metric on $G^-_{r,n}$. The case $r = 1$ gives *complex hyperbolic space.*

The principal bundle $\pi : M^* \to G^-_{r,n}$ has a global holomorphic section since Z_0 is invertible for all planes in $G^-_{r,n}$: If Z is a local holomorphic section, then $W = ZZ_0^{-1}$ does not depend on the choice of Z and $W_0 = 1$, the $r \times r$ unit matrix. The map

$$G^-_{r,n} \to \mathbb{C}^{(n-r)\times r}, \quad [Z] \mapsto w := W_1,$$

is biholomorphic onto the bounded domain $D = \{w \in \mathbb{C}^{(n-r)\times r} \mid \bar{w}^t w < 1\}$. We leave it as an exercise to compute the induced action of $\mathrm{U}(r, n-r)$ on D.

In the case of complex hyperbolic space, that is, if $r = 1$, D is the unit ball in $\mathbb{C}^{n-1}$, and the Kähler form ω on D is given by

$$\begin{aligned} \omega &= -i\partial\bar{\partial}(1 - |w|^2) \\ &= \frac{i}{(1-|w|^2)^2}\Big((1-|w|^2)\sum_j dw^j \wedge d\bar{w}^j + \sum_{j,k} \bar{w}^j w^k dw^j \wedge d\bar{w}^k\Big). \end{aligned} \tag{4.15}$$

For $n = 2$, we get the real hyperbolic plane with curvature -2. As in the previous example, the case $1 < r < n-1$ is more complicated and the explicit formula for ω in the coordinates w does not seem to be of much use.

7) *Bergmann metric.* We refer to [Wei, pages 57–65] for details of the following construction. Let M be a complex manifold of dimension m and $H(M)$ be the space of holomorphic differential forms φ of type $(m, 0)$ on M such that

$$i^{m^2} \int_M \varphi \wedge \bar{\varphi} < \infty.$$

In a first step one shows that $H(M)$ together with the Hermitian product

$$(\varphi, \psi) := i^{m^2} \int_M \varphi \wedge \bar{\psi}$$

is a separable complex Hilbert space. For a unitary basis (φ_n) of $H(D)$, set

$$\theta := i^{m^2} \sum \varphi_n \wedge \bar{\varphi}_n.$$

In a second step one shows that this sum converges uniformly on compact subsets of M and that θ is a real analytic differential form of type (m, m) on M which does not depend on the choice of unitary basis of $H(M)$.

Let $M_0 \subset M$ be the open subset of points $p \in M$ such that there are an open neighborhood U of $p \in M$ and $\varphi_0, \dots, \varphi_m \in H(M)$ such that $\varphi_0(q) \neq 0$ for all $q \in U$ and such that the holomorphic map $z\colon U \to \mathbb{C}^m$ given by $\varphi_j = z^j \varphi_0$ defines a holomorphic coordinate chart of M on U. We may have $M_0 = \emptyset$.

In holomorphic coordinates z on an open subset U of M, write

$$\theta = i^{m^2} f dz^1 \wedge \cdots \wedge dz^m \wedge d\bar{z}^1 \wedge \cdots \wedge d\bar{z}^m.$$

Then f is a non-negative real analytic function on U and $f > 0$ on $U \cap M_0$. The differential form $\omega := i\partial\bar{\partial} \ln f$ on $U \cap M_0$ does not depend on the choice of coordinates, hence ω is a real analytic differential form of type $(1, 1)$ on M_0. The symmetric bilinear form $g(X, Y) := \omega(X, JY)$ is a Kähler metric on M_0 with associated Kähler form ω. By construction and definition, M_0 and θ, hence also ω and g, are invariant under biholomorphic transformations of M. More generally, any biholomorphic transformation $F\colon M \to M'$ induces a biholomorphic transformation between M_0 and M_0' such that the pull back $F^*\theta' = \theta$.

If the group of biholomorphic transformations of M is transitive on M_0, then, by invariance, $\theta = c \operatorname{vol}$ for some constant c, where vol denotes the volume form of g. In holomorphic coordinates z, this reads

$$i^m f dz^1 \wedge d\bar{z}^1 \wedge \cdots \wedge dz^m \wedge d\bar{z}^m = c i^m \det G d\bar{z}^1 \wedge \cdots \wedge dz^m \wedge d\bar{z}^m,$$

where we use notation as in (4.20) below. Hence $f = c \det G$. From (4.63) below we conclude that $g = -\operatorname{Ric}$, where Ric denotes the Ricci tensor of g. Hence g is an Einstein metric with Einstein constant -1 if the group of biholomorphic transformations of M is transitive on M_0.

If D is a bounded domain in $\mathbb{C}^m$, then $\varphi_0 = dz^1 \wedge \cdots \wedge dz^m$ and $\varphi_j = z^j \varphi_0$, $1 \le j \le m$, belong to $H(D)$. Hence $D_0 = D$; we obtain a Kähler metric g on D, the *Bergmann metric* of D.

Let D be as in the previous example. Then up to a scaling factor, the Bergmann metric of D coincides with the Kähler metric defined there since the action of $\mathrm{U}(r, n-r)$ is transitive on D and preserves both metrics and since the induced action of the stabilizer $\mathrm{U}(r) \times \mathrm{U}(n-r)$ of 0 on the tangent space $T_0 D = \mathbb{C}^{(n-r)\times r}$ is irreducible. The scaling factor can be determined by the above observation that the Bergmann metric of D is Einsteinian with Einstein constant -1.

4.16 Proposition. *Let M be a complex manifold with a compatible Riemannian metric $g = \langle \cdot, \cdot \rangle$ as in* (2.33) *and Levi-Civita connection ∇. Then*

$$\begin{aligned} d\omega(X, Y, Z) &= \langle (\nabla_X J)Y, Z\rangle + \langle (\nabla_Y J)Z, X\rangle + \langle (\nabla_Z J)X, Y\rangle, \\ 2\langle (\nabla_X J)Y, Z\rangle &= d\omega(X, Y, Z) - d\omega(X, JY, JZ). \end{aligned}$$

Proof. Since M is a complex manifold, we can assume that the vector fields X, Y, Z, JY, and JZ commute. Then

$$d\omega(X, Y, Z) = X\omega(Y, Z) + Y\omega(Z, X) + Z\omega(X, Y),$$

and similarly for $d\omega(X, JY, JZ)$. The first equation is now immediate from the definition of ω and the characteristic properties of ∇. As for the second, we have

$$\begin{aligned}\langle(\nabla_X J)Y, Z\rangle &= \langle\nabla_X(JY), Z\rangle - \langle J(\nabla_X Y), Z\rangle \\ &= \langle\nabla_X(JY), Z\rangle + \langle\nabla_X Y, JZ\rangle.\end{aligned}$$

By the Koszul formula and the definition of ω,

$$\begin{aligned}2\langle\nabla_X(JY), Z\rangle &= X\langle JY, Z\rangle + JY\langle X, Z\rangle - Z\langle X, JY\rangle \\ &= X\omega(Y, Z) - JY\omega(JZ, X) + Z\omega(X, Y)\end{aligned}$$

and

$$\begin{aligned}2\langle\nabla_X Y, JZ\rangle &= X\langle Y, JZ\rangle + Y\langle X, JZ\rangle - JZ\langle X, Y\rangle \\ &= -X\omega(JY, JZ) + Y\omega(Z, X) - JZ\omega(X, JY),\end{aligned}$$

where we use that X, Y, Z, JY, and JZ commute. □

4.17 Theorem. *Let M be a complex manifold with a compatible Riemannian metric $g = \langle\cdot,\cdot\rangle$ as in* (2.33) *and Levi-Civita connection ∇. Then the following assertions are equivalent:*

1. *g is a Kähler metric.*
2. $d\omega = 0$.
3. $\nabla J = 0$.
4. *In terms of holomorphic coordinates z, we have*
$$\frac{\partial g_{j\bar{k}}}{\partial z^l} = \frac{\partial g_{l\bar{k}}}{\partial z^j} \quad \text{or, equivalently,} \quad \frac{\partial g_{j\bar{k}}}{\partial \bar{z}^l} = \frac{\partial g_{j\bar{l}}}{\partial \bar{z}^k}.$$
5. *The Chern connection of the Hermitian metric h on TM as in Remark* 4.3 *is equal to the Levi-Civita connection ∇.*
6. *For each point p_0 in M, there is a smooth real function f in a neighborhood of p_0 such that $\omega = i\partial\bar{\partial}f$.*
7. *For each point p_0 in M, there are holomorphic coordinates z centered at p_0 such that $g(z) = 1 + O(|z|^2)$.*

A function f as in (6) will be called a *Kähler potential*, holomorphic coordinates as in (7) will be called *normal* coordinates at p_0. The existence of normal coordinates shows that a Kähler manifold agrees with the model example $\mathbb{C}^m$ in (4.10) up to terms of order two and higher.

Proof of Theorem 4.17. By definition, (1) and (2) are equivalent. The equivalence of (2) and (3) is immediate from Proposition 4.16. The equivalence of the two assertions in (4) follows from barring the respective equation. The equivalence of (2) with (4) is immediate from $d = \partial + \bar\partial$ and the formulas defining ∂ and $\bar\partial$. The conclusions (6) $\Rightarrow$ (2) and (7) $\Rightarrow$ (3) are easy and left as an exercise.

We show now that (3) is equivalent with (5). Let X be a local holomorphic vector field and Y be another vector field on M. Then

$$\nabla_{JY}X = \nabla_X(JY) + [JY, X] = \nabla_X(JY) + J[Y, X],$$

where we use that X is automorphic, see Proposition 2.18. On the other hand,

$$J\nabla_Y X = J(\nabla_X Y + [Y, X]) = J\nabla_X Y + J[X, Y].$$

Hence J is parallel iff ∇X is of type (0, 1) for all local holomorphic vector fields of M. It is now obvious that (5) $\Rightarrow$ (3). Vice versa, if J is parallel, then ∇ is Hermitian with respect to h, hence (3) $\Rightarrow$ (5) by what we just said.

We show next that (2) $\Rightarrow$ (6). Since ω is real and $d\omega = 0$, we have $\omega = d\alpha$ locally, where α is a real 1-form. Then $\alpha = \beta + \bar\beta$, where β is a form of type (1,0). Since ω is of type (1,1), we have

$$\partial\beta = 0, \quad \bar\partial\bar\beta = 0 \quad \text{and} \quad \omega = \bar\partial\beta + \partial\bar\beta.$$

Hence $\beta = \partial\varphi$ locally, where φ is a smooth complex function. Then $\bar\beta = \bar\partial\bar\varphi$ and hence

$$\omega = \bar\partial\partial\varphi + \partial\bar\partial\bar\varphi = \partial\bar\partial(\bar\varphi - \varphi) = i\partial\bar\partial f$$

with $f = i(\varphi - \bar\varphi)$.

We explain now that (4) $\Rightarrow$ (7). Let z be holomorphic coordinates centered at p_0 such that $g(0) = 1$. Then $g(z) = 1 + O(|z|)$. We solve for new holomorphic coordinates $\tilde z$ such that

$$z^j := \tilde z^j + \frac{1}{2} A^j_{kl} \tilde z^k \tilde z^l,$$

where

$$A^j_{kl} = -\frac{\partial g_{l\bar j}}{\partial z^k}(0).$$

Applying (4) we get that $\tilde g(\tilde z) = 1 + O(|\tilde z|^2)$. □

4.18 Remarks. 1) Let J be an almost complex structure on a manifold M and g be a compatible Riemannian metric on M. If J is parallel with respect to the Levi-Civita connection of g, then J is a complex structure and hence M is a Kähler manifold, see Exercise 2.6.2. This generalizes Criterion 3 of Theorem 4.17.

2) It is immediate that $\nabla J = 0$ iff $T'M, T''M$ are parallel subbundles of the complexified tangent bundle $T_{\mathbb{C}}M = TM \otimes \mathbb{C}$ with respect to the canonical complex linear extension of ∇ to $T_{\mathbb{C}}M$, $\nabla(X + iY) = \nabla X + i\nabla Y$.

4.1 Kähler Form and Volume. One of the most basic features of a Kähler manifold is the intimate connection between its Kähler form and volume. Let M be a Kähler manifold of complex dimension m with Kähler form ω. There are the following expressions for ω,

$$\omega = i g_{j\bar{k}} dz^j \wedge d\bar{z}^k = \sum X_j^* \wedge Y_j^* = \frac{i}{2} \sum Z_j^* \wedge \bar{Z}_j^*, \tag{4.19}$$

where $(z^1, \dots, z^m)$ are local holomorphic coordinates, $(X_1, Y_1, \dots, X_m, Y_m)$ is a local orthonormal field with $JX_j = Y_j$, and $(Z_j, \bar{Z}_j)$ is the associated complex frame as in (3.30). In turn, we get the following equivalent expansions of ω^m,

$$\begin{aligned} \omega^m &= i^m m! \det G\, dz^1 \wedge d\bar{z}^1 \wedge \cdots \wedge dz^m \wedge d\bar{z}^m \\ &= m! X_1^* \wedge Y_1^* \wedge \cdots \wedge X_m^* \wedge Y_m^* \\ &= i^m 2^{-m} m! Z_1^* \wedge \bar{Z}_1^* \wedge \cdots \wedge Z_m^* \wedge \bar{Z}_m^* \\ &= m! \operatorname{vol}, \end{aligned} \tag{4.20}$$

where $G = (g_{j\bar{k}})$ and where vol denotes the volume form of M.

4.21 Remark. A differential two-form ω on a manifold M is called a *symplectic form* if $d\omega = 0$ and ω is non-degenerate at each point of M, that is, for each point $p \in M$ and non-zero vector $v \in T_pM$ there is a vector $w \in T_pM$ such that $\omega(v, w) \neq 0$. A manifold together with a symplectic form is called a *symplectic manifold.* If ω is a symplectic form on M, then the real dimension of M is even, $\dim M = 2m$, and ω^m is non-zero at each point of M.

The Kähler form of a Kähler manifold is symplectic. Vice versa, the Darboux theorem says that a symplectic manifold (M, ω) admits an atlas of smooth coordinate charts $z \colon U \to U' \subset \mathbb{R}^{2m} \cong \mathbb{C}^m$ such that $z^*\omega_{\text{can}} = \omega$, where

$$\omega_{\text{can}} = \sum dx^j \wedge dy^j = \frac{i}{2} \sum dz^j \wedge d\bar{z}^j$$

is the Kähler form of the Euclidean metric on $\mathbb{C}^m$ as in Example 4.10.1. References for the basic theory of symplectic forms and manifolds are [Ar, Chapter 8] and [Du, Chapter 3]. Compare also our discussion of symplectic Lagrangian spaces in Example B.53.

4.22 Exercise (Compare (5.43)). Show that $* \exp \omega = \exp \omega$, i.e., that

$$* \frac{1}{j!} \omega^j = \frac{1}{(m-j)!} \omega^{m-j}.$$

4.23 Theorem. *Let M be a Kähler manifold as above.*

1) *If M is closed, then the cohomology class of ω^k in $H^{2k}(M,\mathbb{R})$ is non-zero for $0 \le k \le m$. In particular, $H^{2k}(M,\mathbb{R}) \neq 0$ for such k.*

2) *If $N \subset M$ is a compact complex submanifold without boundary of complex dimension k, then the cohomology class of ω^k in $H^{2k}(M,\mathbb{R})$ and the homology class of N in $H_{2k}(M,\mathbb{R})$ are non-zero.*

Proof. Evaluation of ω^m on the fundamental cycle of M, that is, integration of ω^m over M gives

$$\int_M \omega^m = m!\,\mathrm{vol}(M) \neq 0.$$

This shows the first assertion. As for the second, we note that N with the induced metric is a Kähler manifold, and the Kähler form is the restriction of ω to N. Hence integrating ω^k over N gives $k!$ times the volume of N. □

4.24 Example. For $m \ge 2$, the Hopf manifold $S^{2m-1} \times S^1$ has vanishing second cohomology, hence does not carry a Kähler metric.

4.25 Remark. Let (M,ω) be a closed symplectic manifold of (real) dimension $2m$. Since ω is closed and ω^m is non-zero at each point of M, the argument in the proof of Theorem 4.23 applies and shows that $H^{2k}(M,\mathbb{R}) \neq 0$, $0 \le k \le m$. It follows that $S^{2m-1} \times S^1$ does not even carry a symplectic form if $m \ge 2$. It is natural to ask whether there are closed manifolds which carry a symplectic form but not a Kähler metric. The answer is yes, see Example 5.37.2

4.26 Wirtinger Inequality. *Let W be a Hermitian vector space with inner product $\langle v, w\rangle = \mathrm{Re}(v,w)$ and Kähler form $\omega(u,v) = \mathrm{Im}(u,v)$. Let $V \subset W$ be a real linear subspace of even real dimension $2k$ and $(v_1,\dots,v_{2k})$ be a basis of V. Then*

$$|\omega^k(v_1,\dots,v_{2k})| \le k!\,\mathrm{vol}_V(v_1,\dots,v_{2k})$$

with equality iff V is a complex linear subspace of W.

Proof. Observe first that both sides of the asserted inequality are multiplied by $|\det A|$ if we transform $(v_1,\dots,v_{2k})$ with $A \in \mathrm{Gl}(2k,\mathbb{R})$ to another basis of V. Hence we are free to choose a convenient basis of V.

Let $v, w \in W$ be orthonormal unit vectors with respect to $\langle\cdot,\cdot\rangle$. Then $\omega(v,w) = \langle Jv, w\rangle \in [-1,1]$ and $\omega(v,w) = \pm 1$ iff $w = \pm Jv$. This shows the assertion in the case $k=1$. In the general case we observe first that the restriction of ω to V is given by a skew-symmetric endomorphism A of V. Hence there are an orthonormal basis $(v_1,\dots,v_{2k})$ of V and real numbers $a_1,\dots,a_k$ such that

$$A(v_{2j-1}) = a_j v_{2j} \quad \text{and} \quad A(v_{2j}) = -a_j v_{2j-1}.$$

By what we said above, $|a_j| \le 1$ with equality iff $v_{2j} = \pm J v_{2j-1}$. Therefore

$$|\omega^k(v_1,\dots,v_{2k})| = k!|a_1\dots a_k| \le k! = k!\,\mathrm{vol}_V(v_1,\dots,v_{2k})$$

with equality iff V is a complex linear subspace of W. □

4.27 Theorem. *Let M be a Kähler manifold as above. Let $N \subset M$ be a compact complex submanifold of real dimension $2k$ with boundary ∂N (possibly empty). Let $P \subset M$ be an oriented submanifold of dimension $2k$ and boundary $\partial P = \partial N$. If $N - P$ is the boundary of a real singular chain, then* $\operatorname{vol} N \le \operatorname{vol} P$ *with equality if P is also a complex submanifold.*

Proof. Since $N - P$ is the boundary of a real singular chain and ω^k is closed, we have

$$\int_N \omega^k - \int_P \omega^k = \omega^k(N - P) = 0.$$

On the other hand, by the Wirtinger inequality 4.26 we have

$$\int_N \omega^k = \operatorname{vol} N \quad \text{and} \quad \int_P \omega^k \le \operatorname{vol} P,$$

and equality holds iff P is a complex submanifold. □

It follows that a complex submanifold $N \subset M$ is *minimal*, that is, the trace of the second fundamental form S of N vanishes. In fact, Theorem 4.27 implies that N is *pluri-minimal* in the sense that any complex curve in N is minimal in M. This can also be seen by the following computation: Let X, Y be tangent vector fields along N. Since N is complex, JY is tangent to N as well and

$$S(X, JY) = (\nabla_X(JY))^\perp = (J\nabla_X Y)^\perp = J(\nabla_X Y)^\perp = JS(X, Y), \tag{4.28}$$

where we use that the normal bundle of N is invariant under J. It follows that S is complex linear in Y, hence, by symmetry, also in X. Therefore

$$S(X, X) + S(JX, JX) = S(X, X) - S(X, X) = 0,$$

which is what we wanted to show.

4.2 Levi-Civita Connection. Let M be a Kähler manifold. Recall the complex linear extension of ∇ to $T_{\mathbb{C}}M$, $\nabla(X + iY) = \nabla X + i\nabla Y$. We now extend ∇ also complex linearly in the other variable,

$$\nabla_{(U+iV)} Z := \nabla_U Z + i\nabla_V Z. \tag{4.29}$$

The extended connection is symmetric with respect to the complex bilinear extension of the Lie bracket to complex vector fields,

$$\nabla_W Z - \nabla_Z W = [W, Z], \tag{4.30}$$

and metric with respect to the complex bilinear extension of the metric of M to $T_{\mathbb{C}}M$,

$$\nabla_Z \langle U, V\rangle = \langle \nabla_Z U, V\rangle + \langle U, \nabla_Z V\rangle. \tag{4.31}$$

We also have

$$\nabla_{\overline{W}}\overline{Z} = \overline{\nabla_W Z}. \tag{4.32}$$

Since $T'M$ and $T''M$ are complex subbundles of $T_{\mathbb{C}}M$, they stay parallel with respect to the extension. More precisely, since J is parallel,

$$\nabla_{(X+iJX)}(Y+iJY) = (\nabla_X Y + \nabla_Y X) + iJ(\nabla_X Y + \nabla_Y X) \tag{4.33}$$

if $[JX, Y] = 0$. Furthermore, if $[X, Y] = [JX, Y] = 0$, then

$$\nabla_{(X+iJX)}(Y-iJY) = \nabla_{(X-iJX)}(Y+iJY) = 0. \tag{4.34}$$

With respect to local holomorphic coordinates z of M, we let

$$Z_j = \frac{\partial}{\partial z^j} = \frac{1}{2}\left(\frac{\partial}{\partial x^j} - i\frac{\partial}{\partial y^j}\right), \quad \overline{Z}_j = \frac{\partial}{\partial \bar{z}^j} = \frac{1}{2}\left(\frac{\partial}{\partial x^j} + i\frac{\partial}{\partial y^j}\right), \tag{4.35}$$

where $z^j = x^j + iy^j$, $1 \le j \le m$. By (4.34), the mixed covariant derivatives vanish,

$$\nabla_{Z_j}\overline{Z}_k = \nabla_{\overline{Z}_j} Z_k = 0. \tag{4.36}$$

Since $T'M$ and $T''M$ are parallel, we get unmixed Christoffel symbols

$$\nabla_{Z_j} Z_k = \Gamma^l_{jk} Z_l \quad \text{and} \quad \nabla_{\overline{Z}_j}\overline{Z}_k = \Gamma^{\bar{l}}_{\bar{j}\bar{k}}\overline{Z}_l. \tag{4.37}$$

All Christoffel symbols of mixed type vanish. Furthermore, by Equation 4.32 we have

$$\Gamma^{\bar{l}}_{\bar{j}\bar{k}} = \overline{\Gamma^l_{jk}}. \tag{4.38}$$

By (4.31) and (4.36),

$$\Gamma^{\mu}_{jk} g_{\mu\bar{l}} = \langle \nabla_{Z_j} Z_k, \overline{Z}_l\rangle = \frac{\partial g_{k\bar{l}}}{\partial z^j}$$

$$\Gamma^{\bar{\mu}}_{\bar{j}\bar{k}} g_{\bar{\mu}l} = \langle \nabla_{\overline{Z}_j}\overline{Z}_k, Z_l\rangle = \frac{\partial g_{\bar{k}l}}{\partial \bar{z}^j} = \overline{\frac{\partial g_{k\bar{l}}}{\partial z^j}}.$$

Thus the Christoffel symbols are given by

$$\Gamma^l_{jk} = \Gamma^{\mu}_{jk} g_{\mu\bar{\nu}} g^{\bar{\nu}l} = \frac{\partial g_{k\bar{\nu}}}{\partial z^j} g^{\bar{\nu}l}, \tag{4.39}$$

where the coefficients $g^{\bar{k}l}$ denote the entries of the inverse of the fundamental matrix.

4.3 Curvature Tensor. The curvature tensor of the complex bilinear extension of ∇ to $T_{\mathbb{C}}M$ is the complex trilinear extension of the usual curvature tensor R of M to $T_{\mathbb{C}}M$. We keep the notation R for the extension. Then R satisfies the usual symmetries, but now, more generally, for complex vector fields X, Y, Z, U, V,

$$\begin{aligned} R(X,Y)Z &= -R(Y,X)Z, \\ R(X,Y)Z + R(Y,Z)X + R(Z,X)Y &= 0, \end{aligned} \tag{4.40}$$

and

$$\begin{aligned} \langle R(X,Y)U, V\rangle &= -\langle R(X,Y)V, U\rangle, \\ \langle R(X,Y)U, V\rangle &= \langle R(U,V)X, Y\rangle. \end{aligned} \tag{4.41}$$

We also have the reality condition

$$R(\bar X, \bar Y)\bar Z = \overline{R(X,Y)Z}. \tag{4.42}$$

In addition, since J is parallel,

$$R(X,Y)JZ = JR(X,Y)Z. \tag{4.43}$$

By (2.33), (4.41), and (4.43),

$$\langle R(X,Y)JU, JV\rangle = \langle R(JX,JY)U, V\rangle = \langle R(X,Y)U, V\rangle. \tag{4.44}$$

Since $T'M$ and $T''M$ are parallel, we have

$$R(U,V)(T'M) \subseteq T'M \quad \text{and} \quad R(U,V)(T''M) \subseteq T''M. \tag{4.45}$$

With respect to local holomorphic coordinates,

$$R(Z_j, Z_k) = R(\bar Z_j, \bar Z_k) = 0. \tag{4.46}$$

For the mixed terms, we get from (4.45)

$$\begin{aligned} R(Z_j, \bar Z_k)Z_l &= R^{\mu}_{j\bar k l} Z_\mu, \quad & R(\bar Z_j, Z_k)\bar Z_l &= R^{\bar\mu}_{\bar j k \bar l}\bar Z_\mu, \\ R(\bar Z_j, Z_k)Z_l &= R^{\mu}_{\bar j k l} Z_\mu, \quad & R(Z_j, \bar Z_k)\bar Z_l &= R^{\bar\mu}_{j \bar k \bar l}\bar Z_\mu. \end{aligned} \tag{4.47}$$

Equation 4.42 implies that

$$R^{\bar\mu}_{\bar j k \bar l} = \overline{R^{\mu}_{j\bar k l}} \quad \text{and} \quad R^{\bar\mu}_{j\bar k\bar l} = \overline{R^{\mu}_{\bar j k l}}. \tag{4.48}$$

By (4.36), we have

$$R^{\mu}_{j\bar k l} = -\frac{\partial \Gamma^{\mu}_{jl}}{\partial \bar z^k} \quad \text{and} \quad R^{\mu}_{\bar j k l} = \frac{\partial \Gamma^{\mu}_{kl}}{\partial \bar z^j}. \tag{4.49}$$

Recall that the sectional curvature determines the curvature tensor. We define the *holomorphic sectional curvature* of M to be the sectional curvature of the complex lines in TM. If X is a non-zero tangent vector of M, then the complex line spanned by X has (X, JX) as a basis over $\mathbb{R}$ and the corresponding holomorphic sectional curvature is

$$K(X \wedge JX) = \frac{\langle R(X, JX)JX, X\rangle}{\|X\|^4}. \tag{4.50}$$

4.51 Proposition. *The holomorphic sectional curvature determines R.*

Proof. We follow the proof of Proposition IX.7.1 in [KN] and consider the quadri-linear map

$$Q(X, Y, U, V) = \langle R(X, JY)JU, V\rangle + \langle R(X, JU)JY, V\rangle + \langle R(X, JV)JY, U\rangle.$$

It is immediate from the symmetries of R listed above that Q is symmetric. Thus polarization determines Q explicitly from its values

$$Q(X, X, X, X) = 3K(X \wedge JX) \cdot \|X\|^4 \tag{4.52}$$

along the diagonal. Hence it suffices to show that Q determines R. Now

$$Q(X, Y, X, Y) = 2\langle R(X, JY)JY, X\rangle + \langle R(X, JX)JY, Y\rangle.$$

By the second symmetry in (4.40),

$$\begin{aligned}\langle R(X, JX)JY, Y\rangle &= -\langle R(JX, JY)X, Y\rangle - \langle R(JY, X)JX, Y\rangle \\ &= \langle R(X, Y)Y, X\rangle + \langle R(X, JY)JY, X\rangle.\end{aligned}$$

Therefore

$$Q(X, Y, X, Y) = 3\langle R(X, JY)JY, X\rangle + \langle R(X, Y)Y, X\rangle,$$

and hence also

$$Q(X, JY, X, JY) = 3\langle R(X, Y)Y, X\rangle + \langle R(X, JY)JY, X\rangle.$$

In conclusion,

$$3Q(X, JY, X, JY) - Q(X, Y, X, Y) = 8\langle R(X, Y)Y, X\rangle. \tag{4.53}$$

Hence Q determines the sectional curvature, hence R. □

4.54 Exercise. The *bisectional curvature* of X, Y is defined to be $\langle R(X, JX)JY, Y\rangle$. Show that it deserves its name,

$$\langle R(X, JX)JY, Y\rangle = \langle R(X, Y)Y, X\rangle + \langle R(X, JY)JY, X\rangle.$$

4.4 Ricci Tensor. The Ricci tensor is defined by

$$\operatorname{Ric}(X, Y) = \operatorname{tr}(Z \mapsto R(Z, X)Y), \tag{4.55}$$

where we can take the trace of the map on the right hand side as an $\mathbb{R}$-linear endomorphism of TM or as a $\mathbb{C}$-linear endomorphism of $T_{\mathbb{C}}M$. The Ricci tensor is symmetric and

$$\operatorname{Ric}(\bar{X}, \bar{Y}) = \overline{\operatorname{Ric}(X, Y)}. \tag{4.56}$$

The associated symmetric endomorphism field of TM or $T_{\mathbb{C}}M$ is also denoted Ric, that is, we let $\langle \operatorname{Ric} X, Y\rangle := \operatorname{Ric}(X, Y)$.

4.57 Proposition. *In terms of an orthonormal frame* $(X_1, JX_1, \dots, X_m, JX_m)$ *of* M, *the Ricci tensor is given by*

$$\operatorname{Ric} X = \sum R(X_j, JX_j)JX.$$

In particular, $\operatorname{Ric} JX = J \operatorname{Ric} X$.

Proof. We compute

$$\begin{aligned}
\operatorname{Ric}(X, Y) &= \sum \langle R(X_j, X)Y, X_j\rangle + \sum \langle R(JX_j, X)Y, JX_j\rangle \\
&= \sum \langle R(X_j, X)JY, JX_j\rangle - \sum \langle R(JX_j, X)JY, X_j\rangle \\
&= \sum \langle R(X, X_j)JX_j, JY\rangle + \sum \langle R(JX_j, X)X_j, JY\rangle \\
&= -\sum \langle R(X_j, JX_j)X, JY\rangle = \sum \langle R(X_j, JX_j)JX, Y\rangle. \qquad \square
\end{aligned}$$

We conclude that

$$\operatorname{Ric}(JX, JY) = \operatorname{Ric}(X, Y). \tag{4.58}$$

and hence there is an associated real differential form of type $(1, 1)$, the *Ricci form*

$$\rho(X, Y) := \operatorname{Ric}(JX, Y). \tag{4.59}$$

In terms of local holomorphic coordinates,

$$\operatorname{Ric}(Z_j, Z_k) = \operatorname{Ric}(\bar{Z}_j, \bar{Z}_k) = 0, \tag{4.60}$$

by (4.47). For the mixed terms, we have

$$\operatorname{Ric}(Z_j, \bar{Z}_k) = R^{\bar{l}}_{\bar{l}j\bar{k}} = -\frac{\partial \Gamma^{\bar{l}}_{\bar{l}\bar{k}}}{\partial z_j} = \operatorname{Ric}_{j\bar{k}} = \overline{\operatorname{Ric}_{\bar{j}k}} = \overline{\operatorname{Ric}(\bar{Z}_j, Z_k)}. \tag{4.61}$$

With $G = (g_{j\bar{k}})$ we get, by (4.39),

$$\frac{\partial \ln \det G}{\partial z^j} = \operatorname{tr}\left(\frac{\partial G}{\partial z^j} \cdot G^{-1}\right) = \Gamma^k_{jk}. \tag{4.62}$$

Therefore,

$$\operatorname{Ric}_{j\bar{k}} = -\frac{\partial^2 \ln \det G}{\partial z^j \partial \bar{z}^k}. \tag{4.63}$$

For the Ricci form we get

$$\rho = i \operatorname{Ric}_{j\bar{k}} dz^j \wedge d\bar{z}^k = -i\, \partial\bar{\partial} \ln \det G. \tag{4.64}$$

This formula shows that ρ is closed. In fact, by the above computation of the Ricci curvature we have

$$\begin{aligned}\rho(X,Y) &= -\sum \langle R(X,Y)E_j, JE_j\rangle = \sum \langle JR(X,Y)E_j, E_j\rangle \\ &= i\sum \overline{\langle R(X,Y)E_j, E_j\rangle} + i\sum \overline{i\langle JR(X,Y)E_j, E_j\rangle} \\ &= i\sum \overline{h(R(X,Y)E_j, E_j)} = i \operatorname{tr} R(X,Y),\end{aligned} \tag{4.65}$$

where we consider TM with the given complex structure J as a holomorphic vector bundle over M with Hermitian metric $h = g + i\omega$. Now ∇ is a connection on TM considered as a complex vector bundle, namely the Chern connection associated to h. By (A.33) and (A.35), the first Chern form of TM associated to ∇ is given by

$$c_1(TM, \nabla) = \frac{i}{2\pi} \operatorname{tr} R = \frac{1}{2\pi}\rho. \tag{4.66}$$

This shows again that ρ is closed. It also shows that $[\rho/2\pi] = c_1(TM)$ in $H^2(M,\mathbb{R})$ and hence that $[\rho/2\pi]$ lies in the image of $H^2(M,\mathbb{Z}) \to H^2(M,\mathbb{R})$.

The induced curvature on the canonical bundle $K_M = A^{m,0}(M,\mathbb{C})$ is given by

$$\hat{R}(X,Y)(dz_1 \wedge \cdots \wedge dz_m) = i\rho(X,Y) \cdot dz_1 \wedge \cdots \wedge dz_m. \tag{4.67}$$

To prove this, we use that the curvature $\hat{R}$ acts as a derivation on $A^*(M,\mathbb{C})$,

$$\begin{aligned}\hat{R}(X,Y)(dz_1 \wedge \cdots \wedge dz_m) &= \sum dz_1 \wedge \cdots \wedge \hat{R}(X,Y)dz_j \wedge \cdots \wedge dz_m \\ &= -\operatorname{tr} R(X,Y)(dz_1 \wedge \cdots \wedge dz_m).\end{aligned}$$

Now the claim follows from (4.65). We conclude that the first Chern form

$$c_1(K_M, \hat{\nabla}) = -\frac{i}{2\pi} \operatorname{tr} R = -\frac{1}{2\pi}\rho = -c_1(TM, \nabla). \tag{4.68}$$

4.5 Holonomy. Kähler manifolds are characterized by the property that their complex structure is parallel. This links the Kähler condition to holonomy[10].

Suppose that M is a connected Kähler manifold, and let p be a point in M. Then the complex structure J_p turns T_pM into a complex vector space. Since J is parallel,

[10] Excellent references for holonomy are [Bes, Chapter 10], [Br], and [Jo, Chapter 3].

the holonomy group $\operatorname{Hol} M = \operatorname{Hol}_p M$ of M at p preserves J_p. Hence up to a unitary isomorphism of (T_pM, J_p) with $\mathbb{C}^m$, $m = \dim_{\mathbb{C}} M$, the holonomy group $\operatorname{Hol} M$ of M is contained in $\mathrm{U}(m)$. Vice versa, let M be a connected Riemannian manifold of dimension $n = 2m$ and p be a point in M. Suppose that after some identification of T_pM with $\mathbb{R}^{2m}$ and of $\mathbb{R}^{2m}$ with $\mathbb{C}^m$, the holonomy group $\operatorname{Hol}_p M \subset \mathrm{U}(m) \subset \mathrm{SO}(2m)$. Then the complex structure J_p on T_pM induced from the complex structure on $\mathbb{C}^m$ extends to a parallel complex structure J on M, and M together with J is a Kähler manifold. We formulate the result of our discussion in the following somewhat sloppy way.

4.69 Proposition. *A connected Riemannian manifold of real dimension* $2m$ *is a Kähler manifold iff its holonomy group is contained in* $\mathrm{U}(m)$. □

If the holonomy along a closed loop c of M at p is given by A on T_pM, viewed as a complex vector space, then the induced holonomy on the fiber $A_p^{m,0}(M, \mathbb{C})$ of the canonical bundle is given by the complex determinant $\det_{\mathbb{C}} A$. Hence $\operatorname{Hol}(M) \subset \mathrm{SU}(m)$ iff $A^{m,0}(M, \mathbb{C})$ has a parallel complex volume form. It also follows that the reduced holonomy group $\operatorname{Hol}_0(M) \subset \mathrm{SU}(m)$ iff $A^{m,0}(M, \mathbb{C})$ is flat. By (4.67) this holds iff M is Ricci-flat, that is, if $\operatorname{Ric} = 0$.

4.70 Proposition. *A connected Riemannian manifold of real dimension* $2m$ *is a Ricci-flat Kähler manifold iff its reduced holonomy group is contained in* $\mathrm{SU}(m)$. □

4.71 Remark. A connected Riemannian manifold M of real dimension $2m$ with holonomy $\operatorname{Hol}(M) \subset \mathrm{SU}(m)$ is called a *Calabi–Yau manifold*[11]. If the real dimension of M is $4k$, we may also have $\operatorname{Hol}(M) \subset \mathrm{Sp}(k) \subset \mathrm{SU}(m)$, then M is called a *hyper-Kähler manifold.* By Proposition 4.70, Calabi–Yau and hyper-Kähler manifolds are Ricci-flat Kähler manifolds. For examples of Calabi–Yau and hyper-Kähler manifolds, see Chapters 6 and 7 in [Jo].

Let M be a simply connected complete Riemannian manifold. Then we have the de Rham decomposition,

$$M = M_0 \times M_1 \times \cdots \times M_k, \tag{4.72}$$

where M_0 is a Euclidean space and M_i, $i \geq 1$, is a simply connected complete Riemannian manifold with irreducible holonomy[12]. Moreover, this decomposition is unique up to a permutation of the factors M_i, $i \geq 1$. Hence we have, at a point $p = (p_0, p_1, \dots, p_k) \in M$,

$$\operatorname{Hol} M = \operatorname{Hol} M_1 \times \cdots \times \operatorname{Hol} M_k. \tag{4.73}$$

After identifying $T_{p_i} M_i$ with the corresponding subspace of T_pM, we also have

$$T_{p_0} M_0 = \operatorname{Fix} \operatorname{Hol} M, \tag{4.74}$$

[11] There are variations of this definition in the literature.

[12] It is understood that $M_0 = \{0\}$ and $k = 0$ are possible.

the set of vectors fixed by $\operatorname{Hol} M$, and, for each $i \geq 1$,

$$T_{p_0}M_0 + T_{p_i}M_i = \operatorname{Fix} \prod_{j \neq i} \operatorname{Hol} M_j. \tag{4.75}$$

Suppose now in addition that M is a Kähler manifold. Then the complex structure J of M is parallel, and hence $J_p A = A J_p$ for all $A \in \operatorname{Hol} M$. By (4.74), we have $J_p T_{p_0} M_0 = T_{p_0} M_0$, hence M_0 is a complex Euclidean space. From (4.75) we conclude that $J_p T_{p_i} M_i = T_{p_i} M_i$ for all $i \geq 1$. Now the de Rham decomposition as in (4.72) implies the de Rham decomposition for Kähler manifolds.

4.76 De Rham Decomposition of Kähler Manifolds. *Let M be a simply connected complete Kähler manifold. Then*

$$M = M_0 \times M_1 \times \cdots \times M_k,$$

where M_0 is a complex Euclidean space and M_i, $i \geq 1$, is a simply connected complete Kähler manifold with irreducible holonomy.

4.6 Killing Fields. We start with a discussion of Killing fields on Riemannian manifolds.

4.77 Proposition (Yano [Ya], Lichnerowicz [Li1]). *Let M be Riemannian and X be a vector field on M. Then*

$$X \text{ is a Killing field} \implies \operatorname{div} X = 0 \text{ and } \nabla^*\nabla X = \operatorname{Ric} X.$$

If M is closed, then the converse holds true as well.

Proof. If X is a Killing field, then $L_X g = 0$ and hence $\operatorname{div} X = \operatorname{tr} L_X g/2 = 0$, see (1.8). Furthermore, $\nabla^*\nabla X = \operatorname{Ric} X$ by Exercise 1.5.2.

Suppose now that M is closed and let $\xi = X^\flat$. The decomposition of the bilinear form $\hat\nabla\xi$ into symmetric and skew-symmetric part is given by

$$\hat\nabla\xi = (\hat\nabla\xi)^{\mathrm{sym}} + (\hat\nabla\xi)^{\mathrm{skew}} = \frac{1}{2} L_X g + \frac{1}{2} d\xi.$$

Using that $\hat\nabla^* = d^*$ on alternating two-forms, we obtain

$$\hat\nabla^*\hat\nabla\xi = \frac{1}{2}\hat\nabla^*(L_X g) + \frac{1}{2} d^* d\xi.$$

The second ingredient is a simple consequence of the Bochner identity 1.38:

$$\nabla^*\nabla X = \operatorname{Ric} X \iff \hat\nabla^*\hat\nabla\xi = \frac{1}{2}\Delta_d(\xi).$$

Assume now that $\operatorname{div} X = 0$ and $\nabla^*\nabla X = \operatorname{Ric} X$. Since $\operatorname{div} X = -d^*\xi$, we obtain

$$\hat\nabla^* L_X g = d d^*\xi = 0.$$

It follows that $L_X g$ is L^2-perpendicular to im $\hat{\nabla}$. However, adding the term $d\xi$ to $L_X g$, which is pointwise and hence L^2-perpendicular to $L_X g$, we get $2\hat{\nabla}\xi \in \operatorname{im}\hat{\nabla}$. We conclude that $L_X g = 0$ and hence that X is a Killing field. □

Recall that a vector field X on a complex manifold M is automorphic iff $[X, JY] = J[X, Y]$ for all vector fields Y on M. In Proposition 2.18 we showed that X is automorphic iff X is holomorphic.

Suppose now that M is a Kähler manifold, and let X be a vector field on M. Since the Levi-Civita connection is equal to the Chern connection of the holomorphic vector bundle TM with the induced Hermitian metric, X is automorphic iff ∇X is a form of type $(1,0)$ with values in TM. In fact,

$$\begin{aligned} \nabla_{JY} X &= \nabla_X JY - [X, JY] \\ &= J\nabla_X Y - [X, JY] \\ &= J\nabla_Y X + J[X, Y] - [X, JY] \end{aligned} \tag{4.78}$$

for any vector field Y on M. That is, X is automorphic iff $\nabla X \circ J = J \circ \nabla X$.

4.79 Proposition (Lichnerowicz [Li2]). *Let M be a Kähler manifold and X be a vector field on M. Then*

$$X \text{ is automorphic} \implies \nabla^*\nabla X = \operatorname{Ric} X.$$

If M is closed, then the converse holds as well.

Proof. We decompose the one-form ∇X into types,

$$\nabla X = \nabla^{1,0} X + \nabla^{0,1} X = \frac{1}{2}(\nabla X - J \circ \nabla X \circ J) + \frac{1}{2}(\nabla X + J \circ \nabla X \circ J).$$

This decomposition is pointwise and hence L^2-perpendicular. From (4.78) we get that X is automorphic iff $\nabla^{0,1} X = 0$.

The key observation is the following:

$$\nabla^*(J \circ \nabla X \circ J) = J \circ \nabla^*(\nabla X \circ J) = -\operatorname{Ric} X, \tag{4.80}$$

where the first equality holds since J is parallel and the second follows from Proposition 4.57. We conclude that

$$2\nabla^*\nabla^{0,1} X = \nabla^*\nabla X - \operatorname{Ric} X.$$

Hence if X is automorphic, then $\nabla^*\nabla X = \operatorname{Ric} X$.

We now repeat the linear algebra argument from the end of the proof of Proposition 4.77: Suppose that M is closed and that $\nabla^*\nabla X = \operatorname{Ric} X$. Then $\nabla^*\nabla^{0,1} X = 0$, hence $\nabla^{0,1} X$ is L^2-perpendicular to $\operatorname{im}\nabla$. Adding the term $\nabla^{1,0} X$, which is L^2-perpendicular to $\nabla^{0,1} X$, we get $\nabla X \in \operatorname{im}\nabla$. Hence $\nabla^{0,1} X = 0$. □

Comparing Propositions 4.77 and 4.79 we arrive at the final result in this subsection.

4.81 Theorem. *Let M be a closed Kähler manifold. Then Killing fields on M are automorphic. Vice versa, an automorphic field is a Killing field iff it is volume preserving.* □

4.82 Exercise. Find non-automorphic Killing fields on $\mathbb{C}^m$, $m \geq 2$.

4.83 Remark. The description of Killing and automorphic vector fields in terms of the equation $\nabla^*\nabla X = \operatorname{Ric} X$ is important for the global geometry of compact Kähler manifolds with $\operatorname{Ric} \leq 0$ or $\operatorname{Ric} > 0$, compare Chapters 6 and 7.

That Killing fields on a closed Kähler manifold M are automorphic can also be seen in a more conceptual way: A Killing field preserves the metric of M. Hence in order to show that it preserves J as well, it is clearly enough to show that it preserves ω, that is, that $L_X\omega = 0$. But ω is parallel and hence a harmonic form.

Now every diffeomorphism of M acts on $H^*(M,\mathbb{R})$, but isometries of M act on $\mathcal{H}^*(M,\mathbb{R})$ as well. Clearly, for isometries the two actions are compatible with the Hodge isomorphism $\mathcal{H}^*(M,\mathbb{R}) \to H^*(M,\mathbb{R})$. It follows that homotopic isometries induce the same maps on $\mathcal{H}^*(M,\mathbb{R})$, hence connected groups of isometries act trivially on $\mathcal{H}^*(M,\mathbb{R})$. In particular, $L_X\alpha = 0$ for all harmonic forms α and all Killing fields X.

5 Cohomology of Kähler Manifolds

Let M be a Kähler manifold. In our applications to cohomology further on we will assume that M is closed. But for the moment this assumption is not necessary since all of our computations are of a local nature.

Let $E \to M$ be a holomorphic vector bundle over M. Let h be a Hermitian metric on E and D be the corresponding Chern connection. In what follows, we use notation and results introduced in Chapter 1.

We consider some endomorphism fields of $A^*(M, E)$ and their relation with exterior differentiation. To that end we recall that each element α of $A^*(M, E)$ can be written uniquely as a sum $\alpha = \sum \alpha_r$, where α_r has degree r, or as a sum $\alpha = \sum \alpha_{p,q}$, where $\alpha_{p,q}$ is of type (p, q). The corresponding maps $P_r \colon \alpha \mapsto \alpha_r$ and $P_{p,q} \colon \alpha \mapsto \alpha_{p,q}$ define fields of projections of $A^*(M, E)$.

The complex structure J of M acts on $A^*(M, E)$ via pull back. The induced endomorphism field of $A^*(M, E)$ is denoted C. Note that C leaves the type of a form invariant. In terms of the above projection fields, we have

$$C = \sum i^{p-q} P_{p,q}. \tag{5.1}$$

The adjoint operator is

$$C^* = \sum i^{q-p} P_{p,q} = C^{-1}. \tag{5.2}$$

We set $W := (\bar{*} \otimes h^*)(\bar{*} \otimes h)$. Again, W leaves the type of forms invariant. Since the real dimension of M is even, we have

$$W = \sum (-1)^r P_r. \tag{5.3}$$

The *Lefschetz map* is defined by

$$L(\alpha) := \omega \wedge \alpha = \sum X_j^* \wedge Y_j^* \wedge \alpha = \frac{i}{2} \sum Z_j^* \wedge \bar{Z}_j^* \wedge \alpha, \tag{5.4}$$

where ω is the Kähler form of M and (X_j, Y_j) and $(Z_j, \bar{Z}_j)$ are frames as in (3.30). The Lefschetz map raises types of forms by $(1, 1)$. To compute the adjoint operator L^* of L we use the first relation in Exercise 1.26 [13]. Since $Z_j^* = 2\bar{Z}_j^\flat$,

$$L^*\alpha = \sum Y_j \llcorner (X_j \llcorner \alpha) = \frac{2}{i} \sum \bar{Z}_j \llcorner (Z_j \llcorner \alpha). \tag{5.5}$$

We have vanishing commutators

$$[L, C] = [L, W] = [L^*, C] = [L^*, W] = 0, \tag{5.6}$$

by (5.1) and (5.3), where we note that C and W leave the type of forms invariant and that L raises and L^* lowers the type by $(1, 1)$.

[13] The L^2-adjoint of a field of endomorphisms is the field of adjoint endomorphisms so that there is no ambiguity here.

5.7 Remark. The operators C, W, and L and their adjoints are real, that is, also defined on $A^*(M,\mathbb{R})$ and corresponding bundle valued forms $A^*(M,E)$, where E is a real vector bundle, endowed with a Riemannian metric.

5.8 Proposition. $[L^*, L] = \sum (m-r)P_r$.

Proof. By (5.4), (5.5), and Exercise 1.26 (applied to complex tangent vectors),

$$\begin{aligned} LL^*\alpha &= \sum Z_j^* \wedge \bar{Z}_j^* \wedge (\bar{Z}_k \llcorner (Z_k \llcorner \alpha)) \\ &= -\sum_{j\neq k} Z_j^* \wedge (\bar{Z}_k \llcorner (\bar{Z}_j^* \wedge (Z_k \llcorner \alpha))) + \sum_{j=k} Z_j^* \wedge \bar{Z}_j^* \wedge (\bar{Z}_j \llcorner (Z_j \llcorner \alpha)). \end{aligned}$$

Vice versa,

$$\begin{aligned} L^*L\alpha &= \sum \bar{Z}_k \llcorner (Z_k \llcorner (Z_j^* \wedge \bar{Z}_j^* \wedge \alpha)) \\ &= -\sum_{j\neq k} \bar{Z}_k \llcorner (Z_j^* \wedge (Z_k \llcorner (\bar{Z}_j^* \wedge \alpha))) + \sum_{j=k} \bar{Z}_j \llcorner (Z_j \llcorner (Z_j^* \wedge \bar{Z}_j^* \wedge \alpha)). \end{aligned}$$

The first terms on the right hand sides of the above formulas coincide, hence cancel in the commutator $[L^*, L]$. Assume now without loss of generality that $\alpha = Z_I^* \wedge \bar{Z}_J^* \otimes \sigma$, where I and J are multi-indices. Then the second term on the right hand side in the first formula counts the number of j with $j \in I$ and $j \in J$. The second term on the right hand side in the second formula counts the number of j with $j \notin I$ and $j \notin J$. □

We recall now that the standard basis vectors

$$X = \begin{pmatrix} 0 & 1 \\ 0 & 0 \end{pmatrix}, \quad Y = \begin{pmatrix} 0 & 0 \\ 1 & 0 \end{pmatrix}, \quad H = \begin{pmatrix} 1 & 0 \\ 0 & -1 \end{pmatrix} \tag{5.9}$$

of $\mathfrak{sl}_2(\mathbb{C})$ satisfy

$$[X, Y] = H, \quad [H, X] = 2X, \quad [H, Y] = -2Y. \tag{5.10}$$

It follows that

$$X \mapsto L^*, \quad Y \mapsto L, \quad H \mapsto \sum (m-r)P_r \tag{5.11}$$

extends to a representation of $\mathfrak{sl}_2(\mathbb{C})$ on (the fibers of) $A^*(M,E)$ and on $\mathcal{A}^*(M,E)$. Finite dimensional representations of $\mathfrak{sl}_2(\mathbb{C})$ split into the sum of irreducible ones, where the latter are classified by integers $n \geq 0$:

$$V_n = \mathbb{C}v + \mathbb{C}Yv + \cdots + \mathbb{C}Y^n v, \tag{5.12}$$

where $v \in V_n$, $v \neq 0$, is *primitive*, that is, $Xv = 0$, and where $HY^k v = (2k-n)v$, see e.g. Lecture 11 in [FH]. We note that $Hv = -nv$.

5.13 Exercise. Show that on $A^r(M,E)$ and for $s \geq 1$,

$$[L^s, L^*] = s(r-m)L^{s-1}.$$

The tensor field L^* is parallel and has constant rank on $A^r(M, E)$. Hence its kernel $A^r_P(M, E)$ of primitive forms of degree r is a parallel subbundle of $A^r(M, E)$. We denote by $\mathcal{A}^r_P(M, E)$ the space of smooth sections of $A^r_P(M, E)$.

By Proposition 5.8 and what we said about representations of $\mathfrak{sl}_2(\mathbb{C})$, we have $A^r_P(M, E) = 0$ for $r > m$. Furthermore, if $\alpha \in A^r_P(M, E)$, $\alpha \neq 0$, then $L^s\alpha \neq 0$ for $0 \leq s \leq m - r$ and $L^s\alpha = 0$ for $s > m - r$. It follows that the parallel tensor field $L^s\colon A^r(M, E) \to A^{r+2s}(M, E)$ is injective for $0 \leq s \leq m - r$ and surjective for $s \geq m - r$, and similarly for $\mathcal{A}^r(M, E)$. Since any finite dimensional representation of $\mathfrak{sl}_2(\mathbb{C})$ splits into irreducible ones,

$$A^r(M, E) = \oplus_{s\geq 0} L^s A_P^{r-2s}(M, E), \tag{5.14}$$

$$\mathcal{A}^r(M, E) = \oplus_{s\geq 0} L^s \mathcal{A}_P^{r-2s}(M, E), \tag{5.15}$$

the *Lefschetz decompositions* of $A^r(M, E)$ and $\mathcal{A}^r(M, E)$.

Since $\sum(m-r)P_r$ preserves types of forms and since L raises and L^* lowers types by $(1, 1)$, we have refined decompositions,

$$A^{p,q}(M, E) = \oplus_{s\geq 0} L^s A_P^{p-s,q-s}(M, E), \tag{5.16}$$

$$\mathcal{A}^{p,q}(M, E) = \oplus_{s\geq 0} L^s \mathcal{A}_P^{p-s,q-s}(M, E), \tag{5.17}$$

the Lefschetz decompositions of $A^{p,q}(M, E)$ and $\mathcal{A}^{p,q}(M, E)$. It is trivial that

$$A^{p,q}(M, E) = A_P^{p,q}(M, E) \quad \text{and} \quad \mathcal{A}^{p,q}(M, E) = \mathcal{A}_P^{p,q}(M, E) \tag{5.18}$$

if $p = 0$ or if $q = 0$.

5.19 Remark. Since the tensor field L is parallel, the Lefschetz decompositions of $A^r(M, E)$ and $A^{p,q}(M, E)$ are parallel. It follows from Exercise 5.13 that they are orthogonal as well.

5.1 Lefschetz Map and Differentials. We now compute commutators with exterior derivatives. It is immediate from (3.32) and (4.19) that

$$[L, \partial^D] = [L, \bar\partial] = [L, d^D] = 0. \tag{5.20}$$

Hence the commutators of the adjoint operators also vanish,

$$[L^*, (\partial^D)^*] = [L^*, \bar\partial^*] = [L^*, (d^D)^*] = 0. \tag{5.21}$$

5.22 Proposition. *We have*

$$[L, (\partial^D)^*] = i\bar\partial, \quad [L, \bar\partial^*] = -i\partial^D, \quad [L^*, \partial^D] = i\bar\partial^*, \quad [L^*, \bar\partial] = -i(\partial^D)^*.$$

Proof. By (3.33) and (4.19),

$$L(\partial^D)^*\alpha = -\frac{i}{2}\sum_{j,k} Z_j^* \wedge \bar Z_j^* \wedge (2Z_k \llcorner \hat D_{\bar Z_k}\alpha) =$$

$$= -i \sum_{j \neq k} Z_k \llcorner (Z_j^* \wedge \bar{Z}_j^* \wedge \hat{D}_{\bar{Z}_k} \alpha) + i \sum_{j=k} Z_j^* \wedge (Z_j \llcorner (\bar{Z}_j^* \wedge \hat{D}_{\bar{Z}_j} \alpha))$$

and

$$\begin{aligned}
(\partial^D)^* L\alpha &= -\frac{i}{2} \sum_{j,k} 2 Z_k \llcorner \hat{D}_{\bar{Z}_k} (Z_j^* \wedge \bar{Z}_j^* \wedge \alpha) \\
&= -i \sum_{j,k} Z_k \llcorner (Z_j^* \wedge (\bar{Z}_j^* \wedge \hat{D}_{\bar{Z}_k} \alpha)).
\end{aligned}$$

In conclusion,

$$\begin{aligned}
[L, (\partial^D)^*]\alpha &= i \sum_j \{ Z_j^* \wedge (Z_j \llcorner (\bar{Z}_j^* \wedge \hat{D}_{\bar{Z}_j} \alpha)) + Z_j \llcorner (Z_j^* \wedge (\bar{Z}_j^* \wedge \hat{D}_{\bar{Z}_j} \alpha)) \} \\
&= i \sum_j \bar{Z}_j^* \wedge \hat{D}_{\bar{Z}_j} \alpha = i \bar{\partial} \alpha.
\end{aligned}$$

The proof of the second equation $[L, \bar{\partial}^*] = -i\partial^D$ is similar. The remaining two equations are adjoint to the first two. □

We now discuss the three Laplacians Δ_d, $\Delta_{\bar{\partial}}$, and Δ_{∂^D} and their relation with the Lefschetz map.

5.23 Proposition. *We have*

$$[L, \Delta_{\bar{\partial}}] = -i R^D \wedge_\varepsilon \quad \text{and} \quad [L, \Delta_{\partial^D}] = i R^D \wedge_\varepsilon,$$

where $R^D \wedge_\varepsilon$ is the operator sending α to $R^D \wedge_\varepsilon \alpha$.

Proof. From Proposition 3.29 we recall that $R^D \wedge_\varepsilon = \partial^D \bar{\partial} + \bar{\partial} \partial^D$. By Equation 5.20 and Proposition 5.22, we have

$$\begin{aligned}
[L, \Delta_{\bar{\partial}}] &= L \bar{\partial} \bar{\partial}^* + L \bar{\partial}^* \bar{\partial} - \bar{\partial} \bar{\partial}^* L - \bar{\partial}^* \bar{\partial} L \\
&= \bar{\partial} L \bar{\partial}^* + \bar{\partial}^* L \bar{\partial} - i \partial^D \bar{\partial} - \bar{\partial} L \bar{\partial}^* - i \bar{\partial} \partial^D - \bar{\partial}^* L \bar{\partial} \\
&= -i(\partial^D \bar{\partial} + \bar{\partial} \partial^D).
\end{aligned}$$

Similarly,

$$\begin{aligned}
[L, \Delta_{\partial^D}] &= L \partial^D (\partial^D)^* + L (\partial^D)^* \partial^D - \partial^D (\partial^D)^* L - (\partial^D)^* \partial^D L \\
&= \partial^D L (\partial^D)^* + (\partial^D)^* L \partial^D + i \bar{\partial} \partial^D - \partial^D L (\partial^D)^* + i \partial^D \bar{\partial} - (\partial^D)^* L \partial^D \\
&= i(\partial^D \bar{\partial} + \bar{\partial} \partial^D).
\end{aligned}$$

□

5.24 Lemma. *We have*

$$\partial^D \bar{\partial}^* + \bar{\partial}^* \partial^D = \bar{\partial} (\partial^D)^* + (\partial^D)^* \bar{\partial} = 0.$$

Proof. By Proposition 5.22, we have

$$\bar\partial(\partial^D)^* + (\partial^D)^*\bar\partial = i\bar\partial L^*\bar\partial - i\bar\partial\bar\partial L^* + iL^*\bar\partial\bar\partial - i\bar\partial L^*\bar\partial = 0$$

since $\bar\partial\bar\partial = 0$. Now $\partial^D\bar\partial^* + \bar\partial^*\partial^D$ is the adjoint operator of $\bar\partial(\partial^D)^* + (\partial^D)^*\bar\partial$. The lemma follows. □

5.25 Theorem. *We have*

$$\Delta_{\bar\partial} + \Delta_{\partial^D} = \Delta_{d^D} \quad \textit{and} \quad \Delta_{\bar\partial} - \Delta_{\partial^D} = [iR^D\wedge_\varepsilon, L^*].$$

In particular, $[L, \Delta_{d^D}] = 0$ *and* Δ_{d^D} *preserves the type of forms.*

Proof. We compute

$$\begin{aligned}\Delta_{d^D} &= (\partial^D + \bar\partial)((\partial^D)^* + \bar\partial^*) + ((\partial^D)^* + \bar\partial^*)(\partial^D + \bar\partial)\\ &= \Delta_{\partial^D} + \Delta_{\bar\partial} + \partial^D\bar\partial^* + \bar\partial(\partial^D)^* + (\partial^D)^*\bar\partial + \bar\partial^*\partial^D = \Delta_{\partial^D} + \Delta_{\bar\partial},\end{aligned}$$

by (5.24). Now the two last claims follow from Proposition 5.23 and since Δ_{∂^D} and $\Delta_{\bar\partial}$ preserve the type of forms.

To prove the second equality, let α and β be differential forms with values in E, where β has compact support. Then there is no boundary term when considering integration by parts as in the L^2-products below. By Equation 5.20 and Proposition 5.22,

$$\begin{aligned}(i\bar\partial^*\alpha, \bar\partial^*\beta)_2 &= (L^*\partial^D\alpha, \bar\partial^*\beta)_2 - (\partial^D L^*\alpha, \bar\partial^*\beta)_2\\ &= (\bar\partial L^*\partial^D\alpha, \beta)_2 - (\bar\partial\partial^D L^*\alpha, \beta)_2\\ &= (L^*\bar\partial\partial^D\alpha, \beta)_2 + (i(\partial^D)^*\partial^D\alpha, \beta)_2 - (\bar\partial\partial^D L^*\alpha, \beta)_2.\end{aligned}$$

Similarly,

$$\begin{aligned}(i\bar\partial\alpha, \bar\partial\beta)_2 &= -(\bar\partial\alpha, i\bar\partial\beta)_2\\ &= -(\bar\partial\alpha, L(\partial^D)^*\beta)_2 + (\bar\partial\alpha, (\partial^D)^*L\beta)_2\\ &= -(L^*\bar\partial\alpha, (\partial^D)^*\beta)_2 + (L^*\partial^D\bar\partial\alpha, \beta)_2\\ &= -(\bar\partial L^*\alpha, (\partial^D)^*\beta)_2 + (i(\partial^D)^*\alpha, (\partial^D)^*\beta)_2 + (L^*\partial^D\bar\partial\alpha, \beta)_2\\ &= -(\partial^D\bar\partial L^*\alpha, \beta)_2 + (i\partial^D(\partial^D)^*\alpha, \beta)_2 + (L^*\partial^D\bar\partial\alpha, \beta)_2.\end{aligned}$$

The sum of the terms on the left hand side is equal to $(i\Delta_{\bar\partial}\alpha, \beta)_2$, hence the second equality. □

5.26 Corollary. *If D is flat, then* $2\Delta_{\bar\partial} = 2\Delta_{\partial^D} = \Delta_d$.

Since L commutes with Δ_{d^D} and $\Delta_{d^D} = (\Delta_{d^D})^*$, L^* commutes with Δ_{d^D} as well. Hence the representation of $\mathfrak{sl}_2(\mathbb{C})$ on $\mathcal{A}^*(M, E)$ as in (5.11) induces a representation of $\mathfrak{sl}_2(\mathbb{C})$ on the space $\mathcal{H}^*(M,E)$ of Δ_d-harmonic forms. Say that a harmonic form α is *primitive* if $L^*\alpha = 0$, and denote by $\mathcal{H}^r_P(M,E) \subset \mathcal{H}^r(M,E)$ the space of primitive harmonic forms of degree r. A primitive harmonic form is primitive at each point of M. This will be important in some applications further on.

The following results are immediate consequences of the above commutation relations and the Lefschetz decompositions of $\mathcal{A}^r(M,E)$ in (5.15) and $\mathcal{A}^{p,q}(M,E)$ in (5.17).

5.27 Hard Lefschetz Theorem. *The map $L^s\colon \mathcal{H}^r(M,E) \to \mathcal{H}^{r+2s}(M,E)$ is injective for $0 \le s \le m-r$ and surjective for $s \ge m-r$, $s \ge 0$. Furthermore,*

$$\mathcal{H}^r(M,E) = \oplus_{s\ge 0} L^s \mathcal{H}^{r-2s}_P(M,E),$$

the Lefschetz decomposition *of $\mathcal{H}^r(M,E)$.* □

If E is the trivial line bundle with the standard Hermitian metric, then $R^D = 0$ and hence $[iR^D\wedge_\varepsilon, L^*] = 0$. This is the important special case of differential forms with values in $\mathbb{C}$. More generally, if D is flat, that is, if $R^D = 0$, then we also have $[iR^D\wedge_\varepsilon, L^*] = 0$ and, in particular, $2\Delta_{\bar\partial} = 2\Delta_{\partial^D} = \Delta_d$.

5.28 Theorem. *Assume that D is flat. Then*

$$\mathcal{H}^r(M,E) = \oplus_{p+q=r}\mathcal{H}^{p,q}(M,E), \quad \mathcal{H}^{p,q}(M,E) = \oplus_{s\ge 0} L^s \mathcal{H}^{p-s,q-s}_P(M,E),$$

the Hodge *and* Lefschetz decompositions *of $\mathcal{H}^r(M,E)$ and $\mathcal{H}^{p,q}(M,E)$.* □

5.29 Remark. Note that L, L^*, and $\sum(m-r)P_r$ preserve square integrability of differential forms so that Theorems 5.27 and 5.32 also hold for the corresponding spaces of square integrable harmonic forms. Moreover, the decompositions are pointwise and hence L^2-orthogonal.

Suppose that D is flat and that $p+q \le m = \dim_{\mathbb{C}} M$. Then, by Theorem 5.28,

$$\begin{aligned}\mathcal{H}^{p,q}(M,E) &= \oplus_{s\ge 0} L^s \mathcal{H}^{p-s,q-s}_P(M,E)\\ &= \mathcal{H}^{p,q}_P(M,E) \oplus L\mathcal{H}^{p-1,q-1}(M,E).\end{aligned} \tag{5.30}$$

Furthermore, the Lefschetz map $L^s\colon \mathcal{H}^{p-s,q-s}_P(M,\mathbb{C}) \to \mathcal{H}^{p,q}(M,\mathbb{C})$ is injective for all $s \ge 0$, by Theorem 5.27.

5.2 Lefschetz Map and Cohomology. If D is flat, then $d^Dd^D = 0$ and we get the associated cohomology $H^*(M,E)$. If M is closed, then cohomology classes in $H^*(M,E)$ are uniquely represented by harmonic forms, by Hodge theory.

5.31 Hard Lefschetz Theorem. *Assume that M is closed and D is flat. Then the map $L^s\colon H^r(M,E) \to H^{r+2s}(M,E)$ is injective for $0 \le s \le m-r$ and surjective for $s \ge m-r$. Furthermore, we have the Lefschetz decomposition*

$$H^r(M,E) = \oplus_{s\ge 0} L^s H^{r-2s}_P(M,E).$$ □

We recall that Poincaré duality also shows that $H^r(M, \mathbb{C}) \cong H^{2m-r}(M, \mathbb{C})$. However, the isomorphism in Theorem 5.31 is not obtained using the Hodge operator but the $(m-r)$-fold cup product with a cohomology class, the Kähler class. The existence of a cohomology class with this property is a rather special feature of closed Kähler manifolds.

5.32 Theorem. *Assume that M is closed and D is flat. Then we have Hodge and Lefschetz decompositions,*

$$H^r(M, E) = \oplus_{p+q=r} H^{p,q}(M, E), \quad H^{p,q}(M, E) = \oplus_{s\geq 0} L^s H_P^{p-s,q-s}(M, E). \quad \square$$

Suppose that M is closed and D is flat. Let

$$h_P^{p,q}(M, E) := \dim H_P^{p,q}(M, E) = \dim \mathcal{H}_P^{p,q}(M, E). \tag{5.33}$$

If $p + q \leq m$, then by Theorem 5.32,

$$h^{p,q}(M, E) = \sum_{s\geq 0} h_P^{p-s,q-s}(M, E) = h_P^{p,q}(M, E) + h^{p-1,q-1}(M, E), \tag{5.34}$$

see also (5.30) above. Recall that $b_r(M) = \dim H^r(M, \mathbb{C})$ and define Betti numbers $b_r(M, E) := \dim H^r(M, E)$.

5.35 Theorem. *If M is closed and D is flat, then*
1) $b_r(M, E) = \sum_{p+q=r} h^{p,q}(M, E)$;
2) $h^{p,q}(M, E) = h^{m-p,m-q}(M, E^*)$;
3) $h^{q,p}(M, \mathbb{C}) = h^{p,q}(M, \mathbb{C}) = h^{m-p,m-q}(M, \mathbb{C})$;
4) *$b_r(M)$ is even if r is odd,* $b_1(M) = 2h^{1,0}(M, \mathbb{C})$.

In particular, $h^{1,0}(M, \mathbb{C})$ is a topological invariant of M.

Proof. Assertion 1) is immediate from Theorem 5.32. Assertion 2) follows from Serre duality 3.12 as does the second equality in Assertion 3) since the dual bundle of the trivial line bundle is the trivial line bundle. Since conjugation commutes with Δ_d (for complex valued differential forms on any Riemannian manifold), the map

$$\mathcal{H}^{p,q}(M, \mathbb{C}) \to \mathcal{H}^{q,p}(M, \mathbb{C}), \quad \varphi \mapsto \bar{\varphi},$$

is (well defined and) a conjugate linear isomorphism. This shows the first equality in Assertion 3). The remaining assertions are immediate consequences of Assertions 1) and 3). $\square$

5.36 Remarks. 1) The above results motivate a graphic arrangement of the Dolbeault cohomology groups in a diamond with vertices $H^{0,0}(M, E)$ and $H^{m,m}(M, E)$ to the south and north and $H^{m,0}(M, E)$ and $H^{0,m}(M, E)$ to the west and east, respectively. This diamond is called the *Hodge diamond.* The Lefschetz isomorphism $L^r\colon H^r(M, E) \to H^{2m-r}(M, E)$ is compatible with the type decomposition and corresponds to a reflection about the equator of the Hodge diamond. In the case

$E = \mathbb{C} = E^*$, Serre duality corresponds to the reflection about the center. Conjugation corresponds to the reflection about the vertical axis between north and south pole.

2) If M is a closed complex manifold, then Betti numbers and Euler characteristic of M satisfy the *Frölicher relations*

$$b_r(M) \le \sum_{p+q=r} h^{p,q}(M,\mathbb{C}) \quad \text{and} \quad \chi(M) = \sum_{p,q}(-1)^{p+q}h^{p,q}(M,\mathbb{C}),$$

see [GH, page 444] and [Hir, Theorem 15.8.1]. In the Kähler case, these are clear from Theorem 5.35.1.

5.37 Examples. 1) We see again that Hopf manifolds $M = S^{2m-1} \times S^1$ do not carry Kähler metrics for $m \ge 2$ since their first Betti number is odd, $b_1(M) = 1$. It is easy to see that $h^{1,0}(M,\mathbb{C}) = 0$, see Lemma 9.4 in Borel's Appendix to [Hir]. In Theorem 9.5 loc. cit. Borel determines the Dolbeault cohomology of Hopf and Calabi–Eckmann manifolds completely.

2) We come back to the question whether there are closed symplectic manifolds which do not carry Kähler metrics, compare Remark 4.25. A first example of this kind is due to Thurston: Let $H(\mathbb{R})$ be the *Heisenberg group*, that is, the group of (3×3)-matrices of the form

$$A(x,y,z) = \begin{pmatrix} 1 & x & z \\ 0 & 1 & y \\ 0 & 0 & 1 \end{pmatrix}, \qquad x,y,z \in \mathbb{R},$$

and $H(\mathbb{Z})$ be the closed subgroup of matrices in $H(\mathbb{R})$ with $x,y,z \in \mathbb{Z}$. The natural projection $H(\mathbb{R}) \to \mathbb{R}^2$, $A(x,y,z) \to (x,y)$, descends to a projection of the quotients

$$\pi\colon N := H(\mathbb{R})/H(\mathbb{Z}) \to \mathbb{R}^2/\mathbb{Z}^2,$$

a fiber bundle with fiber a circle $\mathbb{R}/\mathbb{Z}$. In particular, N is a closed manifold. Since $H(\mathbb{R})$ is diffeomorphic to $\mathbb{R}^3$, $H(\mathbb{R})$ is simply connected. Therefore the fundamental group of N is isomorphic to $H(\mathbb{Z})$. It follows that the first homology $H_1(N,\mathbb{Z}) = H(\mathbb{Z})/[H(\mathbb{Z}),H(\mathbb{Z})] \cong \mathbb{Z}^2$, and hence $b_1(N) = 2$.

Let $M := N \times S^1$, where the parameter of the factor S^1 is denoted t. In this notation, the differential form $\omega = dx \wedge dt + dy \wedge dz$ is well defined and symplectic on M. However, M does not carry a Kähler metric since the first Betti number of M is 3, an odd number. For more on this topic see [TO] and the survey [BT]. Compare also with the Iwasawa manifold in Example 5.58.

3) There are closed complex manifolds satisfying $h^{p,q}(M,\mathbb{C}) \ne h^{q,p}(M,\mathbb{C})$. For example, the Iwasawa manifold in Example 5.58 satisfies $h^{1,0}(M,\mathbb{C}) = 3$ and $h^{0,1}(M,\mathbb{C}) = 2$, see [Zh, Example 8.9]. This is one of the arguments that show that this manifold does not carry a Kähler metric.

On a closed oriented manifold M of real dimension $2m$, the *intersection form* S is a bilinear form on the middle cohomology,

$$S\colon H^m(M,\mathbb{Z}) \times H^m(M,\mathbb{Z}) \to \mathbb{Z}, \qquad S(x,y) = \langle x \cup y, [M]\rangle, \tag{5.38}$$

where $[M]$ denotes the oriented fundamental cycle of M and the angle brackets denote evaluation. Poincaré duality implies that S is non-degenerate. If m is odd, then S is skew-symmetric. If m is even, then S is symmetric. Thus closed oriented manifolds with real dimension divisible by four come with a remarkable topological invariant, a non-degenerate integral symmetric bilinear form on their middle cohomology.

We can coarsen this picture and consider the intersection form on $H^m(M,\mathbb{R})$. Via representing closed differential forms, it is then given by

$$S(\varphi,\psi)=\int_M \varphi\wedge\psi. \tag{5.39}$$

The index of S is called the *signature* of M, denoted $\sigma(M)$.

In the case of a closed Kähler manifold, there is a variation T of S which is defined on $H^r(M,\mathbb{R})$, $0\le r\le m$. Via representing closed differential forms,

$$T(\varphi,\psi):=\int_M \omega^{m-r}\wedge\varphi\wedge\psi. \tag{5.40}$$

We may also view T as a complex bilinear form on $H^r(M,\mathbb{C})=H^r(M,\mathbb{R})\otimes\mathbb{C}$. We have

$$T(\varphi,\psi)=(-1)^r T(\psi,\varphi) \quad\text{and}\quad T(C\varphi,\psi)=T(\varphi,C^*\psi). \tag{5.41}$$

For the proof of the second equality, we note that since the volume form has type (m,m), we have $T(\varphi,\psi)=0$ if φ is of type (p,q) and ψ is of type (p',q') with $(p,q)\ne(q',p')$.

In our discussion below, we represent cohomology classes of M by their harmonic representatives. It is important that the harmonic representative of a primitive cohomology class is a differential form which is primitive at each point of M.

5.42 Theorem (Hodge–Riemann Bilinear Relations). *Let M be a closed Kähler manifold. Let φ,ψ be primitive cohomology classes of type (p,q) and (p',q') and $s,t\ge 0$ satisfy $p+q+2s=p'+q'+2t=r\le m$. Then*

$$T(L^s\varphi,L^t\psi)=0 \quad \textit{unless } s=t,\ p'=q,\ \textit{and } q'=p; \tag{1}$$

$$i^{q-p}(-1)^{\frac{(p+q)(p+q-1)}{2}}T(L^s\bar\varphi,L^s\varphi)>0 \quad \textit{unless } \varphi=0. \tag{2}$$

Proof. If $s=t$, then (1) is immediate since the volume form has type (m,m). Since φ is primitive, $L^{m-p-q+1}\varphi=0$. Now $m-r+2s=m-p-q$. Hence if $s<t$, then

$$\omega^{m-r}\wedge L^s\varphi\wedge L^t\psi=L^{m-r+s+t}\varphi\wedge\psi=0.$$

The case $s>t$ is similar. This completes the proof of (1).

The main and only point in the proof of (2) is a formula for the $*$-operator on primitive differential forms. If $s\le m-p-q$, then

$$*L^s\varphi=(-1)^{\frac{(p+q)(p+q-1)}{2}}\frac{s!}{(m-p-q-s)!}L^{m-p-q-s}C\varphi. \tag{5.43}$$

For a proof see e.g. [Wei], Théorème 2 on page 23.

Since L^* is a real operator, $\bar\varphi$ is primitive of degree (q,p). Hence, by (5.43),

$$\begin{aligned} i^{q-p}T(L^s\bar\varphi, L^s\varphi) &= \int_M \omega^{m-p-q} \wedge C\bar\varphi \wedge \varphi \\ &= (-1)^{\frac{(p+q)(p+q-1)}{2}}(m-p-q)! \int_M *\bar\varphi \wedge \varphi \\ &= (-1)^{\frac{(p+q)(p+q-1)}{2}}(m-p-q)! \int_M |\varphi|^2. \end{aligned}$$

□

We discuss a few applications of the Hodge–Riemann bilinear relations. Let r be even. Then T is symmetric, see (5.41), hence the form $T_H(\varphi,\psi) = T(\bar\varphi,\psi)$ is Hermitian on $H^r(M,\mathbb{C})$. By Theorem 5.42, T_H is non-degenerate and Hodge and Lefschetz decomposition are orthogonal with respect to T_H.

Let $p,q,s \geq 0$ be given with $p+q+2s = r$. Since r is even, $p+q$ is even as well, that is, p and q have the same parity. Hence $(p+q)^2$ is a multiple of 4 and therefore

$$(-1)^{\frac{(p+q)(p+q-1)}{2}} i^{q-p} = i^{(p+q)(p+q-1)+q-p} = (-1)^p = (-1)^q.$$

From Theorem 5.42.2 we conclude that T_H is positive definite on $L^s\mathcal{H}_P^{p,q}(M,\mathbb{C})$ if p and q are even and negative definite if p and q are odd. For example, since $\dim H^{0,0}(M,\mathbb{C}) = 1$, the index of T_H on $H^{1,1}(M,\mathbb{C})$ is $(1, h^{1,1}(M,\mathbb{C})-1)$, compare also the discussion of Kähler surfaces below.

5.44 Hodge Index Theorem. *Let M be a closed Kähler manifold with even complex dimension m. Then the signature*

$$\sigma(M) = \sum_{p+q \text{ even}} (-1)^q h^{p,q}(M,\mathbb{C}) = \sum_{p,q} (-1)^q h^{p,q}(M,\mathbb{C}).$$

Proof. By Theorem 5.35, $\sum_{p+q=k}(-1)^q h^{p,q}(M,\mathbb{C}) = 0$ if k is odd. Hence we only need to prove the first equality. To that end, we note that by the above discussion

$$\sigma(M) = \sum_{\substack{p+q\leq m \\ p,q \text{ even}}} h_P^{p,q}(M,\mathbb{C}) - \sum_{\substack{p+q\leq m \\ p,q \text{ odd}}} h_P^{p,q}(M,\mathbb{C}).$$

On the other hand, since $p+q \leq m$,

$$h_P^{p,q}(M,\mathbb{C}) = h^{p,q}(M,\mathbb{C}) - h^{p-1,q-1}(M,\mathbb{C}),$$

see (5.34). This gives

$$\sigma(M) = \sum_{\substack{p+q\leq m \\ p+q \text{ even}}} (-1)^q (h^{p,q}(M,\mathbb{C}) - h^{p-1,q-1}(M,\mathbb{C}))$$

The terms $h^{p,q}(M,\mathbb{C})$ with $p+q$ even and $< m$ occur twice and with the same sign. Since $h^{p,q}(M,\mathbb{C}) = h^{m-p,m-q}(M,\mathbb{C})$ and $p = m-p, q = m-q \bmod 2$, we conclude that

$$\sigma(M) = \sum_{p+q \text{ even}} (-1)^q h^{p,q}(M,\mathbb{C}). \qquad \square$$

5.45 Remarks. Suppose that the complex dimension m of M is even.

1) If $p+q = m$ then $p = q$ modulo 2 so that we can substitute $(-1)^p$ for $(-1)^q$ in Theorem 5.44.

2) For p fixed, $\chi_p(M,\mathbb{C}) := \sum_q (-1)^q h^{p,q}(M,\mathbb{C})$ is the Euler characteristic of the cohomology of the Dolbeault chain complex

$$\cdots \xrightarrow{\bar\partial} \mathcal{A}^{p,q-1}(M,\mathbb{C}) \xrightarrow{\bar\partial} \mathcal{A}^{p,q}(M,\mathbb{C}) \xrightarrow{\bar\partial} \mathcal{A}^{p,q+1}(M,\mathbb{C}) \xrightarrow{\bar\partial} \cdots$$

For $p = 0$, we get the arithmetic genus $\chi_0(M,\mathbb{C})$ of M as in (2.31). The Hirzebruch–Riemann–Roch formula expresses the numbers $\chi_p(M,\mathbb{C})$ in terms of the Chern classes of the tangent bundle TM, viewed as a complex vector bundle. We get that the sum of these numbers is a topological invariant of M, $\sum \chi_p(M,\mathbb{C}) = \sigma(M)$.

We point out a few other applications of the Hodge–Riemann bilinear relations 5.42. Let M be a closed Kähler manifold with Kähler form ω. The space $H^1(M,\mathbb{C})$ and the dimensions of $H^{1,0}(M,\mathbb{C})$ and $H^{0,1}(M,\mathbb{C})$ are topological invariants of M, but the inclusions of $H^{1,0}(M,\mathbb{C})$ and $H^{0,1}(M,\mathbb{C})$ into $H^{1,0}(M,\mathbb{C})$ depend on the complex structure of M. By the Hodge–Riemann bilinear relations, the form T is a symplectic form on $H^1(M,\mathbb{C})$ and the subspaces $H^{1,0}(M,\mathbb{C})$ and $H^{0,1}(M,\mathbb{C})$ are Lagrangian subspaces; for the latter, compare Example B.53. The associated Hermitian form $T_H(\varphi,\psi) := iT(\bar\varphi,\psi)$ is negative definite on $H^{1,0}(M,\mathbb{C})$ and positive definite on $H^{0,1}(M,\mathbb{C})$. Thus we obtain a map, the *period map*, which associates to a complex structure on M with a Kähler form cohomologous to ω the Lagrangian subspace $H^{1,0}(M,\mathbb{C})$ in the *period domain*, the Hermitian symmetric space of Lagrangian subspaces on which T_H is negative definite, see again Example B.53. In other words, we obtain a tool to study the space of complex structures on M together with a Kähler structure with Kähler form cohomologous to ω. For example, let M be a closed surface of genus g. Then $H^1(M,\mathbb{C}) \cong \mathbb{C}^{2g}$ and T is the intersection form on $H^1(M,\mathbb{C})$. By the Torelli theorem, the period map is a holomorphic embedding of Teichmüller space into $G^-_{\mathbb{C}}(L,g) = \mathrm{Sp}(g,\mathbb{R})/\mathrm{U}(g)$. Note that by definition, the period map is equivariant with respect to the induced actions of diffeomorphisms.

Suppose now that M is a closed Kähler surface, that is, $\dim_{\mathbb{C}} M = 2$. Then the Hodge numbers $h^{1,0} = h^{1,0}(M,\mathbb{C})$, $h^{0,1}$, $h^{2,1}$, and $h^{1,2}$ are topological invariants of M. We have

$$2h^{2,0} + h^{1,1} = b_2(M) = b_2 \quad \text{and} \quad h^{1,1} = h^{1,1}_P + h^{0,0} = h^{1,1}_P + 1.$$

By the Hodge–Riemann bilinear relations, the Hermitian form $T_H(\varphi,\psi) = S(\bar\varphi,\psi)$ on $H^2(M,\mathbb{C})$ is negative definite on $H^{1,1}_P(M,\mathbb{C})$ and positive definite on the remaining summands in the Hodge decomposition of $H^2(M,\mathbb{C})$. Hence the intersection form S of M has signature $(b_2 - h^{1,1} + 1, h^{1,1} - 1)$ and the signature of M is $b_2 - 2h^{1,1} + 2$. We conclude that the remaining Hodge numbers $h^{2,0}$, $h^{0,2}$, and $h^{1,1}$ are invariants of the smooth structure of M. Furthermore, we obtain again a period map, in this case from the space of complex structures on M together with a Kähler structure with Kähler form cohomologous to ω to the corresponding period domain, the Grassmannian of subspaces of $H^2(M,\mathbb{C})$ of dimension $h_P^{1,1}$ on which T_H is negative definite; for the latter, compare Examples 4.10.6 and B.42. See [Wel, Section V.6], [Jo, Section 7.3], and [CMP] for more on period maps, period domains, and Torelli theorems.

5.3 The dd_c-Lemma and Formality. We assume throughout this subsection that D is flat, that is, that $R^D = 0$. Then the connection form θ of D with respect to a local holomorphic frame is holomorphic, see Proposition 3.21. It follows that E has parallel local holomorphic frames. If $(\Phi_1,\dots,\Phi_k)$ is such a frame and $\sigma = \varphi^\mu\Phi_\mu$ a local smooth section of E, then $d^D\sigma = d\varphi^\mu \otimes \Phi_\mu$ and $\partial^D\sigma = \partial\varphi^\mu \otimes \Phi_\mu$. Therefore we will omit the superindex D and simply write d and ∂ instead of d^D and ∂^D. From Theorem 5.25 we conclude that

$$\Delta_d = 2\Delta_{\bar\partial} = 2\Delta_\partial.$$

We introduce the *complex differential*

$$d_c = i(\bar\partial - \partial) = C^* dC = C^{-1} dC, \tag{5.46}$$

where C is as in (5.1). The two last expressions show that d_c is real, that is, that d_c is well defined on the space $\mathcal{A}^*(M,\mathbb{R})$ of real valued differential forms on M. We have

$$d_c^* = C^* d^* C = C^{-1} d^* C \tag{5.47}$$

and

$$dd_c = 2i\partial\bar\partial = -2i\bar\partial\partial = -d_c d, \tag{5.48}$$

where we use that D is flat. Let $\Delta_{d_c} = d_c d_c^* + d_c^* d_c$ be the Laplacian of d_c. Since M is a Kähler manifold, Δ_d preserves the type of forms. Therefore

$$\Delta_d = C^{-1}\Delta_d C = \Delta_{d_c}. \tag{5.49}$$

5.50 Lemma (dd_c-Lemma). *Suppose that M is closed and D is flat. Let α be a differential form with values in E with*

$$\alpha = d\beta \text{ and } d_c\alpha = 0 \quad \text{or, respectively,} \quad d\alpha = 0 \text{ and } \alpha = d_c\beta.$$

Then there is a differential form γ with values in E such that $\alpha = dd_c\gamma$.

Proof. In what follows, δ denotes one of the operators d or d_c. By (5.49), we have $\Delta_d = \Delta_{d_c} =: \Delta$. Recall the eigenspace decomposition of the Hilbert space of (equivalence classes of) square integrable sections,

$$L^2(A^*(M,E)) = \oplus V_\lambda,$$

where $V_\lambda \subset \mathcal{A}^*(M, E)$ is the eigenspace of Δ for the eigenvalue λ. Since δ commutes with Δ on $\mathcal{A}^*(M, E)$, we have $\delta(V_\lambda) \subset V_\lambda$.

Let $G\colon L^2(A^*(M, E)) \to L^2(A^*(M, E))$ be the Green's operator associated to Δ. Recall that

$$G(\alpha) = \begin{cases} 0 & \text{if } \alpha \in V_0, \\ \lambda^{-1}\alpha & \text{if } \alpha \in V_\lambda \text{ with } \lambda > 0. \end{cases}$$

It follows that G and δ commute on $\mathcal{A}^*(M, E)$. Since G is self-adjoint, we conclude that G and δ^* commute as well on $\mathcal{A}^*(M, E)$. Elliptic regularity implies that $G(\mathcal{A}^*(M, E)) \subset \mathcal{A}^*(M, E)$. This is also immediate from the displayed formula.

The Hodge decomposition theorem asserts that we have an L^2-orthogonal sum

$$\mathcal{A}^*(M, E) = \mathcal{H}^*(M, E) \oplus \delta(\mathcal{A}^*(M, E)) \oplus \delta^*(\mathcal{A}^*(M, E)),$$

where $\mathcal{H}^*(M, E) = V_0 = \ker \Delta$ is the space of δ-harmonic forms. Any differential form β with values in E satisfies

$$\beta = \mathcal{H}\beta + \Delta G\beta,$$

where $\mathcal{H}$ denotes the L^2-orthogonal projection onto $\mathcal{H}^*(M, E)$. By assumption and the Hodge decomposition,

$$\mathcal{H}\alpha = 0$$

since $\alpha \in \operatorname{im}\delta$. The latter also implies $\delta\alpha = 0$, hence

$$\alpha = \Delta G\alpha = (\delta\delta^* + \delta^*\delta)G\alpha = \delta\delta^*G\alpha,$$

where we use that G and δ commute on $\mathcal{A}^*(M, E)$. In the first case we also assume that $d_c\alpha = 0$, in the second that $d\alpha = 0$, hence

$$\alpha = d_c d_c^* G\alpha \quad \text{and} \quad \alpha = dd^*G\alpha$$

in the first and second case, respectively. In both cases we conclude

$$\alpha = dd^*Gd_cd_c^*G\alpha = dd^*d_cGd_c^*G\alpha = -dd_c(d^*Gd_c^*G\alpha),$$

where we note that the commutation relation 5.24 implies that $d^*d_c = -d_cd^*$. □

5.51 Exercises. 1) Calculate that, for functions φ,

$$\langle dd_c\varphi, \omega\rangle = 2\langle i\partial\bar\partial\varphi, \omega\rangle = -2g^{j\bar k}\frac{\partial^2\varphi}{\partial z_j\partial \bar z_k} = \Delta\varphi.$$

This is a truly remarkable formula: It shows that, for Kähler manifolds, the Laplacian of a function does not involve derivatives of the metric. The formula implies that, for M closed and connected, the kernel of $dd_c\colon \mathcal{A}^0(M, \mathbb{C}) \to \mathcal{A}^2(M, \mathbb{C})$ consists precisely of the constant functions.

2) If α is a differential form of type (p, q) and $\alpha = d\beta$ or $\alpha = d_c\beta$, then $\alpha = dd_c\gamma$.

3) Formulate and prove a $\partial\bar\partial$-Lemma.

What follows is taken from [DGMS]. Assume throughout that M is closed and D is flat. Since $dd_c = -d_c d$, the subcomplex $\mathcal{A}^*_{d_c}(M, E)$ of $\mathcal{A}^*(M, E)$ consisting of d_c-closed differential forms is closed under the exterior differential d. Let $H^*_{d_c}(M, E)$ be the cohomology of the complex $(\mathcal{A}^*_{d_c}(M, E), d)$.

5.52 Lemma. *The inclusion $\mathcal{A}^*_{d_c}(M, E) \to \mathcal{A}^*(M, E)$ induces an isomorphism*

$$H^*_{d_c}(M, E) \xrightarrow{\cong} H^*(M, E).$$

Proof. We show first that the induced map on cohomology is surjective. Let α be a closed differential form on M. Let $\beta = d_c\alpha$. Then $d\beta = -d_c d\alpha = 0$, hence $\beta = dd_c\gamma$, by Lemma 5.50, and

$$d_c(\alpha + d\gamma) = \beta + d_c d\gamma = \beta - dd_c\gamma = 0.$$

Hence the cohomology class of α contains a representative in $\mathcal{A}^*_{d_c}(M, E)$.

Suppose now that $d_c\alpha = 0$ and $\alpha = d\beta$. Then $\alpha = d(d_c\gamma)$, by Lemma 5.50. Now $d_c(d_c\gamma) = 0$, hence α is a d-coboundary in $\mathcal{A}^*_{d_c}(M, E)$, and hence the induced map on cohomology is injective. □

Set now $\bar{\mathcal{A}}^*_{d_c}(M, E) := \mathcal{A}^*_{d_c}(M, E)/d_c\mathcal{A}^*(M, E)$. Then d induces a differential $\bar{d}$ on $\bar{\mathcal{A}}^*_{d_c}(M, E)$ and turns the latter into a chain complex. The cohomology of the complex $(\bar{\mathcal{A}}^*_{d_c}(M, E), \bar{d})$ will be denoted $\bar{H}^*_{d_c}(M, E)$.

5.53 Lemma. *The differential $\bar{d} = 0$ and the projection $\mathcal{A}^*_{d_c}(M, E) \to \bar{\mathcal{A}}^*_{d_c}(M, E)$ induces an isomorphism*

$$H^*_{d_c}(M, E) \xrightarrow{\cong} \bar{H}^*_{d_c}(M, E).$$

Proof. Let $\alpha \in \mathcal{A}^*_{d_c}(M, E)$ and set $\beta = d\alpha$. Then $d_c\beta = -dd_c\alpha = 0$, and hence $d\alpha = \beta = d_c d\gamma \in \operatorname{im} d_c$, by Lemma 5.50. Hence $\bar{d} = 0$.

Let $\alpha \in \mathcal{A}^*_{d_c}(M, E)$. Then $d_c\alpha = 0$ and hence $d\alpha = d_c d\beta$ for some β. Let $\gamma = \alpha + d_c\beta$. Then $d_c\gamma = 0$, hence $\gamma \in \mathcal{A}^*_{d_c}(M, E)$, $d\gamma = 0$, hence γ represents an element in $H^*_{d_c}(M, E)$, and finally $\alpha = \gamma$ modulo $d_c\mathcal{A}^*(M, E)$.

Let now $\alpha \in \mathcal{A}^*_{d_c}(M, E)$ with $d\alpha = 0$ and such that $\alpha = d_c\beta$. Then $\alpha = dd_c\gamma$ for some γ, hence α is cohomologous to 0 in $\mathcal{A}^*_{d_c}(M, E)$. □

For E the trivial line bundle, the chain complex $(\mathcal{A}_{d_c}(M, \mathbb{C}), d)$ is invariant under the wedge product and the chain complex $(\bar{\mathcal{A}}^*(M, \mathbb{C}), \bar{d})$ therefore also inherits a product, turning them into graded differential algebras over $\mathbb{C}$, commutative in the graded sense as in the case of differential forms. We use cgda as a shorthand for commutative graded differential algebra.

A homomorphism between cgda's A and B (over a field F) is called a *quasi-isomorphism* if it induces an isomorphism of their cohomology rings. Then A and B

are called quasi-isomorphic, in letters $A \approx B$. A cgda A is called *formal* if there is a sequence $B_0 = A, B_1, \dots, B_k$ of cgda's over F such that $B_{i-1} \approx B_i$, $1 \le i \le k$, and such that the differential of B_k is trivial. For a closed Kähler manifold M we have established that

$$\mathcal{A}^*(M, \mathbb{C}) \leftarrow \mathcal{A}^*_{d_c}(M, \mathbb{C}) \to \mathcal{A}^*_{d_c}(M, \mathbb{C})/d_c\mathcal{A}^*(M, \mathbb{C}) \tag{5.54}$$

are quasi-isomorphisms and that the differential of the latter is trivial.

5.55 Theorem ([DGMS]). *If M is a closed Kähler manifold, then M is formal over $\mathbb{C}$, that is, $\mathcal{A}^*(M, \mathbb{C})$ is a formal commutative differential graded algebra.* □

Since d_c is a real operator, the proofs also work in the case of real numbers and show that $\mathcal{A}^*(M, \mathbb{R})$ is a formal cgda. In other words, M is also formal over $\mathbb{R}$.

5.56 Remark (See Remark 1.42). Let M be a closed Riemannian manifold with curvature operator $\hat{R} \ge 0$. Then real valued harmonic forms on M are parallel. Since the wedge product of parallel differential forms is parallel, hence harmonic, we get that $\mathcal{H}^*(M, \mathbb{R})$ together with wedge product and trivial differential is a cgda over $\mathbb{R}$. Moreover, the inclusion

$$\mathcal{H}^*(M, \mathbb{R}) \to \mathcal{A}^*(M, \mathbb{R})$$

induces an isomorphism in cohomology, see (1.36). Hence M is formal over $\mathbb{R}$.

We refer to [DGMS] for a discussion of the meaning of formality for the topology of M. In a sense made precise in Theorem 3.3 and Corollary 3.4 in [DGMS], the $\mathbb{R}$-homotopy type of a closed and simply connected Kähler manifold M is a formal consequence of the real cohomology ring of M.

To give at least one application of Theorem 5.55, we explain one of the features which distinguish formal cgda's from general ones. Let A be a cgda and $a, b, c \in H^*(A)$ be cohomology classes of degree p, q and r, respectively, such that $a \cup b = b \cup c = 0$. Let $\alpha, \beta, \gamma \in A$ be representatives of a, b, c, respectively. Then $\alpha \wedge \beta = d\varphi$ and $\beta \wedge \gamma = d\psi$ for $\varphi, \psi \in A$ of degree $p + q - 1$ and $q + r - 1$, respectively. The *Massey triple product*

$$\langle a, b, c \rangle := [\varphi \wedge \gamma - (-1)^p \alpha \wedge \psi] \in H^{p+q+r-1}(A). \tag{5.57}$$

We leave it as an exercise to the reader to show that the Massey triple product is well defined up to $\alpha \wedge H^{q+r-1}(A) + \gamma \wedge H^{p+q-1}(A)$, that it is preserved under quasi-isomorphisms, and that it vanishes for cgda's with trivial differential. We conclude that Massey triple products vanish for formal cgda's, in particular for the real cohomology of closed Kähler manifolds.

5.58 Example. Let $G = H(\mathbb{C})$ be the *complex Heisenberg group*, that is, the complex Lie group of matrices $A(x, y, z)$ as in Example 5.37.2, but now with x, y, z in $\mathbb{C}$. Let $\Gamma \subset H(\mathbb{C})$ be the discrete subgroup of $A(x, y, z)$ with x, y, z in $R := \mathbb{Z} + i\mathbb{Z}$. Since Γ acts by biholomorphic maps on G, the *Iwasawa manifold* $M = \Gamma \backslash G$, is a complex manifold of complex dimension 3. The homomorphism $G \to \mathbb{C}^2$, $A(x, y, z) \mapsto (x, y)$, induces a holomorphic submersion $\pi : M \to R^2\backslash\mathbb{C}^2$ with fibers biholomorphic to $R\backslash\mathbb{C}$. In particular, M is a closed manifold and, more specifically, a torus bundle over a torus.

The differential $(1, 0)$-forms dx, dy, and $dz - xdy$ on G are left-invariant, hence they descend to differential $(1, 0)$-forms a, b, and φ on M. We have $da = db = 0$ and $d\varphi = -a \wedge b$. Let $\alpha = [a]$, $\beta = [b]$ in $H^1(M, \mathbb{C})$. Since π induces an isomorphism from $H_1(M, \mathbb{Z}) = \Gamma/[\Gamma, \Gamma] = R^2$ to $H_1(R^2\backslash\mathbb{C}^2, \mathbb{Z}) = R^2$, we conclude that $(\alpha, \bar{\alpha}, \beta, \bar{\beta})$ is a basis of $H^1(M, \mathbb{C})$. We have

$$\langle \alpha, \beta, \beta \rangle = [\varphi \wedge b].$$

Since $d\varphi \neq 0$, it follows that $\langle \alpha, \beta, \beta \rangle \neq 0$ modulo $\alpha \wedge H^1(M, \mathbb{C}) + \beta \wedge H^1(M, \mathbb{C})$. It follows that the Iwasawa manifold is not formal over $\mathbb{C}$. Hence M does not carry a Kähler metric[14]. Note that the first Betti number of M is 4. It can be shown that $b_2(M) = 8$ and $b_3(M) = 10$, so that the Betti numbers of M alone do not show that M is not Kählerian.

A *nilmanifold* is a quotient $\Gamma\backslash G$, where G is a nilpotent Lie group and Γ is a discrete subgroup of G. The Iwasawa manifold M in Example 5.58 is a nilmanifold since $H(\mathbb{C})$ is nilpotent. A theorem of Benson–Gordon and Hasegawa says that a compact nilmanifold $\Gamma\backslash G$ does not admit a Kähler structure unless G is Abelian, see [BeG], [Ha], [TO]. This generalizes our discussion in Example 5.58.

5.4 Some Vanishing Theorems. Let M be a closed complex manifold. We say that a differential form ω on M of type $(1, 1)$ is *positive* and write $\omega > 0$ if

$$g(X, Y) := \omega(X, JY) \tag{5.59}$$

is a Riemannian metric on M. If ω is closed and positive, then g turns M into a Kähler manifold with Kähler form ω. We say that a cohomology class $c \in H^{1,1}(M, \mathbb{C})$ is *positive* and write $c > 0$ if c has a positive representative. We say that a holomorphic line bundle $E \to M$ is *positive*, denoted $E > 0$, if its first Chern class $c_1(E) > 0$. Negative differential forms and cohomology classes of type $(1, 1)$ and negative line bundles are defined correspondingly. We have $E > 0$ iff the dual bundle $E^* < 0$.

Let $E \to M$ be a holomorphic line bundle over M. Let h be a Hermitian metric on E and D be the associated Chern connection. We note that the curvature form Θ of D acts by scalar multiplication on sections and is independent of the local trivialization. Thus Θ and R^D are complex valued differential forms on M. In this sense we have $\Theta = R^D$.

5.60 Examples. We continue the discussion of Examples 3.45 and consider some line bundles over $\mathbb{C}P^m$. As before, we will use the open covering of $\mathbb{C}P^m$ by the sets $U_j = \{[z] \in \mathbb{C}P^m \mid z_j \neq 0\}$, $0 \leq j \leq m$.

[14]For interesting other arguments, see [GH, page 444] and [Zh, Example 8.9]. See also [FG].

1) The tautological bundle $U \to \mathbb{C}P^m$. In the notation of Example 3.45.1, the canonical Hermitian metric h on U is given by

$$h(\varphi_j, \varphi_j) = \frac{1}{z_j \bar{z}_j} \sum z_k \bar{z}_k.$$

In the point $z \in U_j$ with coordinates $z_j = 1$ and $z_k = 0$ for $k \neq j$, the curvature of the Chern connection D of h is

$$\Theta = \bar{\partial}\partial \ln h(\varphi_j, \varphi_j) = \sum_{k \neq j} \bar{\partial}\partial(z_k \bar{z}_k) = -\sum_{k \neq j} dz_k \wedge d\bar{z}_k,$$

as we already checked in Example 3.24 in greater generality. Hence, in the above point z,

$$c_1(U, D) = -\frac{i}{2\pi} \sum_{k \neq j} dz_k \wedge d\bar{z}_k < 0.$$

Now the homogeneity of U and h under the canonical action of the unitary group $\mathrm{U}(m+1)$ implies that $c_1(U, D) < 0$ everywhere. Hence U is negative.

2) The canonical line bundle $K = A^{m,0}(\mathbb{C}P^m, \mathbb{C}) \to \mathbb{C}P^m$: We saw in Example 3.45.2 that K is isomorphic to U^{m+1}. For the curvature of the Chern connection of the canonical metric on K we get $(m+1)\Theta$, where Θ is the curvature of U with respect to the Chern connection of its canonical Hermitian metric. By the previous example, we have $i(m+1)\Theta < 0$, hence $K < 0$.

More generally, for a Kähler manifold M with Ricci form ρ, the curvature of the canonical bundle K_M is given by $i\rho$, see (4.67). Since the canonical metric on $\mathbb{C}P^m$ has positive Ricci curvature, this confirms $i(m+1)\Theta = -\rho < 0$.

3) The hyperplane bundle $H \to \mathbb{C}P^m$: H is inverse to U, that is, H is isomorphic to the dual line bundle U^*. It follows that H is positive.

5.61 Lemma. *Let $E \to M$ be a holomorphic line bundle. Then $E > 0$ iff there is a Hermitian metric h on E such that the curvature Θ of its Chern connection satisfies $i\Theta > 0$, and similarly in the case $E < 0$.*

Proof. Since $2\pi c_1(E) = [i\Theta]$, E is positive if E has a Hermitian metric with $i\Theta > 0$. Suppose now that $E > 0$. Choose a Kähler metric g on M with Kähler form $\omega \in 2\pi c_1(E)$. This is possible since $c_1(E) > 0$. Let h be any Hermitian metric on E, Θ be the curvature form of its Chern connection, and φ be a local and nowhere vanishing holomorphic section of E (a local frame of E). Then

$$\Theta = \bar{\partial}\partial \ln h_\varphi,$$

where $h_\varphi = h(\varphi, \varphi)$. In particular, $\bar{\partial}\Theta = \partial\Theta = 0$. We also have $\bar{\partial}\omega = \partial\omega = 0$. Now $[i\Theta] = 2\pi c_1(E) = [\omega]$ by the definition of $c_1(M)$ and the choice of ω. Since M is a closed Kähler manifold, the dd_c-Lemma 5.50 applies and shows that

$$\omega - i\Theta = i\partial\bar{\partial}\psi$$

for some real function ψ. Let $h' = h \cdot e^{\psi}$. Then the curvature Θ' of the Chern connection of h' satisfies

$$i\Theta' = i\bar{\partial}\partial \ln h' = i\Theta + \omega - i\Theta = \omega > 0. \qquad \square$$

5.62 Proposition. *Let $E \to M$ be a holomorphic line bundle and h be a Hermitian metric on M such that the curvature Θ of its Chern connection D satisfies $i\lambda\Theta > 0$ for some $\lambda \in \mathbb{R}$. Then $\omega := i\lambda\Theta$ is the Kähler form of a Kähler metric on M, and with respect to this Kähler metric*

$$\Delta_{\bar{\partial}} - \Delta_{\partial^D} = \frac{1}{\lambda} \sum (r - m) P_r.$$

Proof. By Theorem 5.25 and Proposition 5.8, we have

$$\Delta_{\bar{\partial}} - \Delta_{\partial^D} = [iR^D \wedge_\varepsilon, L^*] = \frac{1}{\lambda}[\omega\wedge, L^*] = \frac{1}{\lambda}[L, L^*] = \frac{1}{\lambda} \sum (r - m) P_r. \qquad \square$$

5.63 Remark. Proposition 5.62 refines the Nakano inequalities, which usually go into the proof of Kodaira's vanishing theorem 5.64, compare [Wei, Proposition VI.2.5]. We note that these inequalities are special cases of the last two (groups of) equations in the proof of Theorem 5.25. In [Mo], Moroianu proves Proposition 5.62 by a direct computation using a Weitzenböck formula.

5.64 Kodaira Vanishing Theorem. *Let M be a closed complex manifold and $E \to M$ be a holomorphic line bundle.*

1. *If $E > 0$, then $H^{p,q}(M, E) = 0$ for $p + q > m$.*
2. *If $E < 0$, then $H^{p,q}(M, E) = 0$ for $p + q < m$.*

Proof. In view of Serre duality 3.12, the two assertions are equivalent since $E > 0$ iff $E^* < 0$. We discuss the second assertion (and a similar argument shows the first). Let h be a Hermitian metric on E such that the curvature Θ of its Chern connection satisfies $i\Theta < 0$. Let g be the Kähler metric on M with Kähler form $\omega = -i\Theta$. Let α be a $\Delta_{\bar{\partial}}$-harmonic form with values in E and of type (p, q). Then, by Proposition 5.62,

$$0 \le (m - p - q)||\alpha||_2^2 = -(\Delta_{\partial^D}\alpha, \alpha)_2 \le 0. \qquad \square$$

5.65 Corollary. *Let M be a closed complex manifold and $E \to M$ be a holomorphic line bundle with $K^* \otimes E > 0$. Then $H^{0,q}(M, E) = 0$ for $q \ge 1$.*

This will be important in the proof of the Kodaira embedding theorem 9.6. Another application: If $K^* \otimes E > 0$, then

$$\chi_0(M, E) := \sum (-1)^q \dim H^{0,q}(M, E) = \dim H^{0,0}(M, E),$$

the dimension of the space of holomorphic sections of E. Hence if $K^* \otimes E > 0$, then the Hirzebruch–Riemann–Roch formula [LM, Theorem III.13.15] computes the dimension of $H^{0,0}(M, E)$.

Proof of Corollary 5.65. We note first that a differential form of type (m,q) with values in a holomorphic bundle L is the same as a form of type $(0,q)$ in the bundle $K \otimes L$. Therefore,

$$H^{0,q}(M,E) = H^{0,q}(M, K \otimes K^* \otimes E) = H^{m,q}(M, K^* \otimes E) = 0$$

by Theorem 5.64 since $m+q > m$. □

Let M be a Kähler manifold and $E \to M$ be a holomorphic line bundle. Let h be a Hermitian metric on E and D be the corresponding Chern connection. Let $p \in M$. Then there exist an orthonormal frame $(X_1, Y_1, \dots, X_m, Y_m)$ with $Y_j = JX_j$ in (or about) p and real numbers $\theta_1, \dots, \theta_m$ such that the curvature of D at p is given by

$$R_p^D = \frac{1}{2} \sum \theta_j Z_j^* \wedge \bar{Z}_j^*, \tag{5.66}$$

where the frame $(Z_1, \bar{Z}_1, \dots, Z_m, \bar{Z}_m)$ is as in (3.30) and $(Z_1^*, \bar{Z}_1^*, \dots, Z_m^*, \bar{Z}_m^*)$ is the corresponding dual frame. For a local section σ of E about p and a differential form $\alpha = Z_J^* \wedge \bar{Z}_K \otimes \sigma$, we have

$$\begin{aligned} [Z_j^* \wedge \bar{Z}_j^* \wedge, L^*]\alpha &= \sum_k \{Z_j^* \wedge (\bar{Z}_j^* \wedge (\bar{Z}_k \llcorner (Z_k \llcorner \alpha))) - \bar{Z}_k \llcorner (Z_k \llcorner (Z_j^* \wedge (\bar{Z}_j^* \wedge \alpha)))\} \\ &= Z_j^* \wedge (\bar{Z}_j^* \wedge (\bar{Z}_j \llcorner (Z_j \llcorner \alpha))) - \bar{Z}_j \llcorner (Z_j \llcorner (Z_j^* \wedge (\bar{Z}_j^* \wedge \alpha))) \\ &= \begin{cases} \alpha & \text{if } j \in J \cap K; \\ -\alpha & \text{if } j \notin J \cup K; \\ 0 & \text{otherwise.} \end{cases} \end{aligned} \tag{5.67}$$

It follows that for α as above

$$[iR_p^D \wedge_\varepsilon, L^*]\alpha = \frac{1}{2}\Big(\sum_{j \in J \cap K} \theta_j - \sum_{j \notin J \cup K} \theta_j \Big)\alpha. \tag{5.68}$$

This refines Proposition 5.8.

We interpret (the proof of) Theorem 5.64 as the case of "constant curvature" $R^D = \mp i\omega$. Theorem 5.25 gives us some room for extensions to pinching conditions. We prove a simple model result in this direction.

5.69 Theorem. *Let M be a closed Kähler manifold with Kähler form ω. Let $E \to M$ be a holomorphic line bundle and h be a Hermitian metric on E with Chern connection D. Let $\kappa \colon M \to \mathbb{R}$ and $\varepsilon > 0$ and suppose that*

$$|R^D(X,Y) + i\kappa\omega(X,Y)| < \varepsilon |X|\,|Y|$$

for all vector fields X and Y. Then $H^{p,q}(M,E) = 0$ for all p, q with

$$\varepsilon \cdot (m - |p-q|) < \kappa \cdot (p+q-m).$$

We note that there are two cases: $\kappa > 0$ for $p+q > m$ and $\kappa < 0$ for $p+q < m$.

Proof of Theorem 5.69. Let $p_0 \in M$. The condition on R^D implies that the numbers θ_j as in (5.66) satisfy $|\theta_j - \kappa(p_0)| < \varepsilon$ for all j. Let I, J be multi-indices with $|I| = p$ and $|J| = q$, and set $a = |I \cap J|$. Then

$$\Big(\sum_{j\in J\cap K} \theta_j - \sum_{j\notin J\cup K} \theta_j\Big) \geq a(\kappa(p_0) - \varepsilon) - (m-p-q+a)(\kappa(p_0)+\varepsilon)$$
$$\geq (p+q-m)\kappa(p_0) - (m-|p-q|)\varepsilon > 0.$$

Hence $([iR^D, L^*]\alpha, \alpha)_2 > 0$ unless $\alpha = 0$. The rest of the argument is as in Theorem 5.64. □

For more on vanishing theorems, see [Ko2] and [ShS].

6 Ricci Curvature and Global Structure

Let M be a closed Riemannian manifold. Recall that the isometry group of M is a compact Lie group (with respect to the compact-open topology) with Lie algebra isomorphic to the Lie algebra of Killing fields on M (with sign of the Lie bracket reversed).

6.1 Theorem (Bochner [Boc]). *Let M be a closed Riemannian manifold with* $\mathrm{Ric} \le 0$ *and X be a vector field on M. Then X is a Killing field iff X is parallel.*

Proof. It is clear that parallel fields are Killing fields. Vice versa, let X be a Killing field on M. By Proposition 4.77, we have $\nabla^*\nabla X = \mathrm{Ric}\, X$. Therefore

$$0 \le \int |\nabla X|^2 = \int \langle \nabla^*\nabla X, X\rangle = \int \langle \mathrm{Ric}\, X, X\rangle \le 0,$$

and hence X is parallel. □

6.2 Corollary. *Let M be a closed and connected Riemannian manifold with* $\mathrm{Ric} \le 0$. *Then the identity component of the isometry group of M is a torus of dimension $k \le$* $\dim M$, *and its orbits foliate M into a parallel family of flat tori. If M is simply connected or if* $\mathrm{Ric} < 0$ *at some point of M, then the isometry group of M is finite.*

Proof. The first assertion is clear from Theorem 6.1 since the Lie algebra of the isometry group is given by the Lie algebra of Killing fields. As for the second assertion, it suffices to show that M does not carry a non-trivial Killing field. Let X be such a field. By Theorem 6.1, X is parallel. If M is simply connected, then X points in the direction of a Euclidean factor $\mathbb{R}$ of M, by the de Rham decomposition theorem. This is in contradiction to the compactness of M. In any case we note that $R(\cdot, X)X = 0$ since X is parallel. This implies $\mathrm{Ric}\, X = 0$ and is in contradiction to $\mathrm{Ric} < 0$. □

What we need in the proof of Theorem 6.1 is the inequality $\mathrm{Ric} \le 0$ together with the differential equation $\nabla^*\nabla X = \mathrm{Ric}\, X$. Recall now that automorphic vector fields on closed Kähler manifolds are characterized by the latter equation, see Proposition 4.79. Hence we have the following analogue of Theorem 6.1.

6.3 Theorem. *Let M be a closed Kähler manifold with* $\mathrm{Ric} \le 0$ *and X be a vector field on M. Then X is automorphic iff X is parallel.* □

We recall now again that the automorphism group of a closed complex manifold M is a complex Lie group with respect to the compact-open topology with Lie algebra isomorphic to the Lie algebra of automorphic vector fields on M (again and for the same reason with sign of the Lie bracket reversed), see [BM].

6.4 Corollary. *Let M be a closed Kähler manifold with* $\mathrm{Ric} \le 0$. *Then the Lie algebras of parallel, Killing, and automorphic vector fields coincide. The identity components of the automorphism group and of the isometry group of M coincide and are complex tori of complex dimension* $\le \dim_{\mathbb{C}} M$. □

6.5 Remark. By a famous result of Lohkamp [Loh], all smooth manifolds of dimension ≥ 3 admit complete Riemannian metrics of strictly negative Ricci curvature. Corollary 6.4 shows that, on closed complex manifolds, there are obstructions against Kähler metrics of non-positive Ricci curvature.

6.1 Ricci-Flat Kähler Manifolds. A celebrated theorem of Cheeger and Gromoll says that the universal covering space $\tilde{M}$ of a closed and connected Riemannian manifold M with $\mathrm{Ric} \geq 0$ splits isometrically as $\tilde{M} = E \times N$, where E is a Euclidean space and N is closed and simply connected with $\mathrm{Ric} \geq 0$, see [CG1]. The following application is given in [Bea].

6.6 Theorem. *Let M be a closed and connected Kähler manifold with* $\mathrm{Ric} = 0$. *Then there is a finite covering $\hat{M}$ of M which splits isometrically as $\hat{M} = F \times N$, where F is a closed flat Kähler manifold and N is a closed and simply connected Kähler manifold with* $\mathrm{Ric} = 0$.

6.7 Remark. By the Bieberbach theorem, a closed flat Riemannian manifold is finitely covered by a flat torus [Wo, Chapter 3].

Proof of Theorem 6.6. According to the de Rham decomposition theorem for Kähler manifolds 4.76 and the theorem of Cheeger and Gromoll mentioned above, the universal covering space $\tilde{M}$ of M decomposes as

$$\tilde{M} = \mathbb{C}^k \times N, \quad k \geq 0,$$

where N is a closed and simply connected Kähler manifold with $\mathrm{Ric} = 0$. Hence the isometry group $I(\tilde{M})$ of $\tilde{M}$ is the product of the isometry group of $\mathbb{C}^k$ and the isometry group $I(N)$ of N. By Corollary 6.2, $I(N)$ is finite.

The fundamental group Γ of M acts isometrically as a group of covering transformations on $\tilde{M}$, hence is a subgroup of $I(\tilde{M})$. Consider the projection of Γ to the factor $I(N)$ of $I(\tilde{M})$. Since $I(N)$ is finite, the kernel $\hat{\Gamma}$ has finite index in Γ and gives a finite cover of M as asserted. □

6.8 Remark. Let (M, J) be a closed complex manifold with $c_1(M) = 0$. Then for each Kähler form ω of M, there is exactly one Ricci-flat Kähler metric g' on M with Kähler form ω' cohomologous to ω, by the Calabi conjecture 7.1. Therefore the space of Ricci-flat Kähler metrics on M can be identified with the cone of Kähler forms in $H^{1,1}(M, \mathbb{R}) = H^{1,1}(M, \mathbb{C}) \cap H^2(M, \mathbb{R})$, an open subset of $H^{1,1}(M, \mathbb{R})$, hence of real dimension $h^{1,1}(M, \mathbb{C})$ if non-empty [15]. A reference for this are Subsection 4.9.2 and Section 6.8 in [Jo], where Joyce discusses deformations of complex and Calabi–Yau manifolds.

[15] We do not elaborate on the topology of spaces of metrics.

6.2 Nonnegative Ricci Curvature. Let M be a closed Kähler manifold and φ be a form of type $(p, 0)$. Let $(X_1, Y_1, \dots, X_m, Y_m)$ be a local orthonormal frame of M such that $JX_j = Y_j$, and define Z_j and $\bar{Z}_j$ as in (3.30). Using that φ is of type $(p, 0)$ and the formulas (3.32) for $\bar{\partial}$ and (3.33) for $\bar{\partial}^*$, we get

$$\Delta_d \varphi = 2\Delta_{\bar{\partial}} \varphi = 2\bar{\partial}^* \bar{\partial} \varphi = -4 \sum \bar{Z}_j \llcorner \bar{Z}_k^* \wedge \hat{D}^2 \varphi(Z_j, \bar{Z}_k).$$

Now $D^2\varphi(Z_j, \bar{Z}_k)$ is still of type $(p, 0)$, hence the right hand side is equal to

$$-4 \sum \hat{D}^2 \varphi(Z_j, \bar{Z}_j) = -\operatorname{tr} \hat{D}^2 \varphi - i \sum \hat{R}(X_j, Y_j)\varphi,$$

where we use Exercise 3.37.2 for the latter equality. In conclusion,

$$\Delta_d \varphi = 2\bar{\partial}^* \bar{\partial} \varphi = \hat{\nabla}^* \hat{\nabla} \varphi + K\varphi, \tag{6.9}$$

with curvature term $K = -i \sum \hat{R}(X_j, Y_j)$. Now the curvature acts as a derivation on $A^*(M, \mathbb{C})$. Hence to compute its action on $A^{p,0}(M, \mathbb{C})$, $p > 0$, we first determine its action on forms of type $(1, 0)$. For $X \in T''M$, the dual vector $X^\flat = \langle X, \cdot \rangle$ is of type $(1, 0)$. For $Y \in T'M$, we get

$$\begin{aligned} \langle KX^\flat, Y^\flat \rangle &= -i \sum \langle \hat{R}(X_j, Y_j) X^\flat, Y^\flat \rangle = -i \sum \langle R(X_j, Y_j) X, Y \rangle \\ &= -\sum \langle R(X_j, Y_j) X, JY \rangle = \operatorname{Ric}(X, Y), \end{aligned}$$

where we recall that $\langle \cdot, \cdot \rangle$ denotes the complex bilinear extension of the Riemannian metric of M to the complexified tangent bundle $T_{\mathbb{C}}M$. Now at a given point in M, we may choose the above frame $(X_1, Y_1, \dots, X_m, Y_m)$ to consist of eigenvectors of the Ricci tensor, considered as an endomorphism. For multi-indices $J\colon j_1 < \dots < j_p$ and $K\colon k_1 < \dots < k_p$, we then have, at the chosen point,

$$\langle K(Z_{j_1}^* \wedge \dots \wedge Z_{j_p}^*), \bar{Z}_{k_1}^* \wedge \dots \wedge \bar{Z}_{k_p}^* \rangle = \begin{cases} 4^p \sum \operatorname{Ric}(\bar{Z}_{j_\mu}, Z_{j_\mu}) & \text{if } J = K, \\ 0 & \text{otherwise,} \end{cases} \tag{6.10}$$

see (3.31). Hence the curvature term K in the above equation is non-negative if Ric is non-negative and positive if Ric is positive.

6.11 Theorem (Bochner [Boc]). *Let M be a closed, connected Kähler manifold with* $\operatorname{Ric} \geq 0$. *Then holomorphic forms of type $(p, 0)$ are parallel; in particular,*

$$h^{p,0}(M, \mathbb{C}) = h^{0,p}(M, \mathbb{C}) \leq \binom{m}{p}.$$

If $\operatorname{Ric} > 0$ *at some point of M, then M does not have non-trivial holomorphic forms of type $(p, 0)$ for $p > 0$, and then $h^{p,0}(M, \mathbb{C}) = h^{0,p}(M, \mathbb{C}) = 0$ for $p > 0$.*

Proof. If φ is a holomorphic form of type $(p,0)$, $p>0$, then $\Delta_d\varphi = 2\bar\partial^*\bar\partial\varphi = 0$. Hence

$$0 = \int_M \langle \hat\nabla^*\hat\nabla\varphi, \varphi\rangle + \int_M \langle K\varphi, \varphi\rangle = \int_M |\hat\nabla\varphi|^2 + \int_M \langle K\varphi,\varphi\rangle.$$

As we just explained, $\langle K\varphi,\varphi\rangle \geq 0$ if $\mathrm{Ric} \geq 0$. □

The last assertion in Theorem 6.11 also follows from the Kodaira vanishing theorem. To see this we note that $\mathrm{Ric} > 0$ implies that the anti-canonical bundle $K_M^* \to M$ is positive, see (4.68). On the other hand, if K_M^* is positive and $E \to M$ is the trivial complex line bundle or, generalizing Definition 5.59 slightly, a non-negative line bundle, then $K_M^* \otimes E$ is positive, and we may apply Corollary 5.65 to conclude that $h^{0,q}(M,E) = 0$ for $q \geq 1$. Observe that Bochner's argument above works in this case as well (Exercise).

6.12 Theorem (Kobayashi [Ko1]). *Let M be a closed, connected Kähler manifold with* $\mathrm{Ric} > 0$. *Then M is simply connected and the arithmetic genus*

$$\chi(M,\mathcal{O}) = \sum(-1)^p h^{0,p}(M,\mathbb{C}) = 1.$$

Proof. Consider the universal covering $\pi\colon \tilde M \to M$. With respect to the induced Kähler structure on $\tilde M$, π is a local isometry. Since the Ricci curvature of M is positive, the fundamental group of M is finite, by the theorem of Bonnet–Myers. Hence $\tilde M$ is closed as well. Now the Hirzebruch–Riemann–Roch formula expresses the arithmetic genus of a closed Kähler manifold as an integral over a universal polynomial in the curvature tensor. Since π is a local isometry, this implies

$$\chi(\tilde M,\mathcal{O}) = |\Gamma|\cdot\chi(M,\mathcal{O}),$$

where Γ denotes the fundamental group of M. By Theorem 6.11,

$$\chi(\tilde M,\mathcal{O}) = \chi(M,\mathcal{O}) = 1,$$

hence $|\Gamma| = 1$, and hence M is simply connected. □

In his article [Ko1], Kobayashi observed the application of the Kodaira vanishing theorem mentioned before Theorem 6.12 and conjectured that closed and connected Kähler manifolds with positive first Chern class are simply connected, compare (4.68). Now the Calabi conjecture implies that closed Kähler manifolds with positive first Chern class carry Kähler metrics with positive Ricci curvature, see Corollary 7.3. Hence they are in fact simply connected, by Kobayashi's original Theorem 6.12.

6.3 Ricci Curvature and Laplace Operator. Let M be a closed, connected Riemannian manifold of real dimension n. Denote by Δ the Laplace operator on functions on M. If $\mathrm{Ric} \geq \lambda > 0$, then the first non-zero eigenvalue λ_1 of Δ satisfies

$$\lambda_1 \geq \frac{n}{n-1}\lambda. \tag{6.13}$$

Moreover, equality holds iff M is a round sphere of radius $\sqrt{(n-1)/\lambda}$. The inequality (6.13) is due to Lichnerowicz [Li3], the equality discussion to Obata [Ob]; see also [BGM, Section III.D]. The next theorem deals with a remarkable improvement of the above inequality in the Kähler case.

6.14 Theorem (Lichnerowicz). *Let M be a closed Kähler manifold with* $\operatorname{Ric} \geq \lambda > 0$. *Then the first non-zero eigenvalue λ_1 of Δ satisfies*

$$\lambda_1 \geq 2\lambda.$$

Equality implies that the gradient field $X = \operatorname{grad}\varphi$ of any eigenfunction φ for λ_1 is automorphic with $\operatorname{Ric} X = \lambda X$.

Proof. Let $\varphi \in \mathcal{E}(M,\mathbb{R})$ be any non-constant eigenfunction, so that $\Delta\varphi = \mu\varphi$ for some $\mu > 0$. Then $\xi := d\varphi$ satisfies $\Delta_d \xi = \mu\xi$. Hence, by the Bochner identity 1.38, $X := \xi^\sharp = \operatorname{grad}\varphi \neq 0$ fulfills

$$\nabla^*\nabla X = (\mu - 2\lambda)X + (2\lambda X - \operatorname{Ric} X).$$

In the notation introduced there, (4.80) implies

$$\nabla^*\nabla^{0,1} X = \frac{1}{2}(\mu - 2\lambda)X + (\lambda X - \operatorname{Ric} X).$$

Therefore, by the assumption $\operatorname{Ric} \geq \lambda$,

$$0 \leq \|\nabla^{0,1}X\|_2^2 = \langle \nabla^{0,1}X, \nabla X\rangle_2 = \langle \nabla^*\nabla^{0,1}X, X\rangle_2 \leq \frac{1}{2}(\mu - 2\lambda)\|X\|_2^2,$$

where the index 2 indicates L^2-inner products and norms. We conclude that $\mu \geq 2\lambda$ for all non-trivial eigenvalues μ of Δ. Equality implies that $\nabla^{0,1}X = 0$, that is, X is automorphic, and that $\operatorname{Ric} X = \lambda X$. □

6.15 Exercise. Discuss Theorem 6.14 in the case of the round S^2 (and observe that $n/(n-1) < 2$ if $n > 2$). Compare with Theorem 6.16.

Let M be a closed Kähler manifold. Let $\mathfrak{a}$ be the complex Lie algebra of automorphic vector fields on M and $\mathfrak{k} \subset \mathfrak{a}$ be the real Lie subalgebra of Killing fields. The next theorem completes the equality discussion in Theorem 6.14 in the case where M is a Kähler–Einstein manifold.

6.16 Theorem (Matsushima). *Let M be a closed Kähler–Einstein manifold with Einstein constant $\lambda > 0$. Then we have an isomorphism*

$$\operatorname{grad}\colon \{\varphi \in \mathcal{E}(M,\mathbb{R}) \mid \Delta\varphi = 2\lambda\varphi\} =: E_{2\lambda} \to J\mathfrak{k}, \quad \varphi \mapsto \operatorname{grad}\varphi.$$

Proof. We note first that grad is injective on the space of smooth functions on M with mean zero.

By the proof of Theorem 7.43 below, $J\mathfrak{k}$ consists precisely of automorphic gradient vector fields. Hence by the equality case in Theorem 6.14, grad maps $E_{2\lambda}$ to $J\mathfrak{k}$. By the above remark, grad is injective on $E_{2\lambda}$.

Vice versa, let X in $J\mathfrak{k}$. Write $X = \operatorname{grad}\varphi$, where φ is a smooth function on M with mean zero. Now X is automorphic, hence $\nabla^*\nabla X = \operatorname{Ric} X = \lambda X$, by Proposition 4.79. Hence $\Delta\varphi = 2\lambda\varphi$, by Exercise 6.17 and since φ and $\Delta\varphi$ have mean zero. □

6.17 Exercise (Ricci Equation). Let M be a Riemannian manifold and φ be a smooth function on M. Show that

$$\operatorname{grad}\Delta\varphi = (\nabla^*\nabla)(\operatorname{grad}\varphi) + \operatorname{Ric}\operatorname{grad}\varphi.$$

We recall that, on functions, $\Delta = \nabla^*\nabla$, where we write ∇ instead of grad. Thus the formula computes the commutator $[\nabla, \nabla^*\nabla]$. Derive an analogous formula for metric connections on vector bundles.

7 Calabi Conjecture

In this section we discuss the Calabi conjecture and present a major part of its proof [16]. We let M be a closed and connected Kähler manifold with complex structure J, Riemannian metric g, Kähler form ω, and Ricci form ρ. Recall that the Ricci form of any Kähler metric on M is contained in $2\pi c_1(M)$.

7.1 Theorem (Calabi–Yau). *Let $\rho' \in 2\pi c_1(M)$ be a closed real $(1,1)$-form. Then there is a unique Kähler metric g' on M with Kähler form ω' cohomologous to ω and with Ricci form ρ'.*

This was conjectured by Calabi, who also proved uniqueness of the solution g' [Ca1], [Ca2]. Existence was proved by Yau [Ya2], [Ya3], [Ka1]. One of the most striking immediate applications of Theorem 7.1 concerns the existence of Ricci-flat Kähler metrics:

7.2 Corollary. *If $c_1(M) = 0$, then M has a unique Ricci-flat Kähler metric g' with Kähler form ω' cohomologous to ω.*

As a consequence, M is finitely covered by a product of a simply connected closed complex manifold with vanishing first Chern class and a complex torus (where one of the factors might be of dimension 0), see Theorem 6.6.

Recall Definition 5.59 of positive and negative cohomology classes of type $(1,1)$. Say that a closed complex manifold M is a *Fano manifold* if $c_1(M)$ is positive or, equivalently, if the anti-canonical bundle K_M^* is ample[17]. Since their first Chern class is represented by their Ricci form, closed Kähler manifolds with positive Ricci curvature are Fano manifolds. For example, Hermitian symmetric spaces of compact type are Fano manifolds. Theorem 7.1 implies that Fano manifolds are characterized by positive Ricci curvature:

7.3 Corollary. *If M is a Fano manifold, then M carries Kähler metrics with positive Ricci curvature.*

Proof. By assumption, M has a positive closed differential form ω of type $(1,1)$. Then $g(X,Y) = \omega(X, JY)$ is a Kähler metric on M with Kähler form ω. Choose a further positive differential form $\rho' \in 2\pi c_1(M)$, for example $\rho' = \omega$. By Theorem 7.1, there is a unique Kähler metric g' with Kähler form cohomologous to ω and Ricci form ρ'. Hence the Ricci curvature of g' is positive. □

The argument for Corollary 7.3 works equally well in the case $c_1(M) < 0$. In this case, however, we have the stronger Theorem 7.14 below.

7.4 Examples. 1) Let $M \subset \mathbb{C}P^n$ be the (regular) set of zeros of a homogeneous polynomial of degree d. Then $c_1(M) = (n+1-d)c$, where c is the restriction of the first Chern class of the hyperplane bundle H over $\mathbb{C}P^n$ to M, see [Hir, p.159].

---

[16] My main sources for this are [Au3] and [Jo]. Other good references are [Bo1], [Si1], [Ti3].

[17] For the equivalence, see Exercise 9.5 and the Kodaira Embedding Theorem 9.6.

By Example 5.60.3, $c > 0$. We conclude that $c_1(M) > 0$ if $d \leq n$, $c_1(M) = 0$ if $d = n+1$, and $c_1(M) < 0$ if $d > n+1$.

2) Let M_k be the blow up of $\mathbb{C}P^2$ in $k \geq 0$ points in (very) general position. Then $c_1(M_k) > 0$ for $0 \leq k \leq 8$. Moreover, besides $\mathbb{C}P^1 \times \mathbb{C}P^1$, these are the only closed complex surfaces with positive first Chern class [Hit], [Ya1]. The rich geometry of these surfaces is discussed in [Va, Lectures 13–15].

From these examples one gets the impression that Fano manifolds might be rare and special. In fact, from Corollary 7.3 and Theorems 6.11 and 6.12 we conclude that Fano manifolds are simply-connected and that they do not admit any holomorphic p-forms for $p > 0$. Moreover, for each $m \geq 1$, there are only finitely many diffeomorphism types of Fano manifolds of complex dimension m, for example 104 for $m = 3$, see [De].

Before we go into the details of the proof of Theorem 7.1, we reformulate it into a more convenient analytic problem and discuss another question, the existence of Kähler–Einstein metrics.

Recall that $\omega^m = m! \operatorname{vol}_g$, where vol_g denotes the oriented volume form of g, see (4.20). Hence if ω' is cohomologous to ω, then g' and g have the same volume. For such an ω' given, there is a smooth function $f \colon M \to \mathbb{R}$ with

$$(\omega')^m = e^f \omega^m. \tag{7.5}$$

Then the volume constraint turns into

$$\int e^f d \operatorname{vol}_g = \operatorname{vol}(M, g). \tag{7.6}$$

In other words, e^f has mean 1. With (4.64) we get

$$\rho' = -i\partial\bar{\partial} \ln \det G' = -i\partial\bar{\partial} \ln(e^f \det G) = \rho - i\partial\bar{\partial} f. \tag{7.7}$$

Vice versa, for any closed real $(1,1)$-form ρ' cohomologous to ρ, there is a smooth function $f \colon M \to \mathbb{R}$ such that

$$\rho' - \rho = -\frac{1}{2} dd_c f = -i\partial\bar{\partial} f, \tag{7.8}$$

by the dd_c-Lemma 5.50 and Exercise 5.51. Moreover, f is unique up to an additive constant. Thus in our search for the Kähler form ω' we have replaced the Ricci form ρ', which depends on second derivatives of the metric, by the function f, which depends only on ω' itself.

For any Kähler form ω' cohomologous to ω, there is a smooth real function φ such that

$$\omega' - \omega = \frac{1}{2} dd_c \varphi = i\partial\bar{\partial}\varphi, \tag{7.9}$$

again by the dd_c-Lemma 5.50. Vice versa, any such form ω' is cohomologous to ω. Thus we seek φ with

$$e^f \omega^m = (\omega')^m = (\omega + i\partial\bar{\partial}\varphi)^m \tag{7.10}$$

and

$$\omega' = \omega + i\partial\bar{\partial}\varphi > 0. \tag{7.11}$$

By the latter we mean that $g'(X, Y) = \omega'(X, JY)$ is a Kähler metric on M.

In a first step we show that $\omega' > 0$ if (7.10) holds. Consider the Hermitian metric on $T'M$ given by

$$(Z, W) = \langle Z, \overline{W} \rangle,$$

where we note that this Hermitian metric is conjugate linear in the second variable. With respect to a coordinate frame $(Z_1, \dots, Z_m)$ of $T'M$ as in (2.13), its fundamental matrix is $(g_{j\bar{k}})$. We write the Hermitian metric $(\cdot, \cdot)'$ on $T'M$ induced by g' as $(\cdot, \cdot)' = (A\cdot, \cdot)$, where A is a Hermitian field of endomorphisms of $T'M$. Since e^f is non-zero everywhere, (7.10) and (4.20) imply that $\det A = e^f \neq 0$ at each point of M. We want to show that at each point of M all eigenvalues of A are positive. To that end we note that with respect to local coordinates

$$M(\varphi) := e^f = \det A = \det(g_{j\bar{k}})^{-1} \det\left(g_{j\bar{k}} + \frac{\partial^2\varphi}{\partial z_j \partial \bar{z}_k}\right). \tag{7.12}$$

It follows that all eigenvalues of A are positive at a point where φ attains a minimum. Such points exist since M is compact. Suppose now that there is a point $p \in M$ such that A has a negative eigenvalue at p. Then on a path from p to a point where φ attains a minimum, there is a point where A has 0 as an eigenvalue. In such a point, $\det A = 0$ in contradiction to (7.10). We have achieved the following reformulation of Theorem 7.1:

Given a smooth function f on M such that e^f has mean 1, there is a smooth real function φ on M such that

$$\ln M(\varphi) = f. \tag{7.13}$$

This is a non-linear partial differential equation of Monge–Ampère type. The solution φ of (7.13) is unique up to constant functions, see Proposition 7.25. Since $\omega' = \omega + i\partial\bar{\partial}\varphi$ is cohomologous to ω, the assumption on the mean of e^f is automatically fulfilled. In other words, we may delete the normalizing assumption on the mean and ask for a solution of (7.13) up to an additive constant $f + c$ instead. This point of view will be useful further on.

Before we discuss the proof of Assertion 7.13, we discuss a related problem, namely the question of the existence of Kähler–Einstein metrics.

7.14 Theorem (Aubin–Calabi–Yau). *Let M be a closed complex manifold with negative first Chern class. Then up to a scaling constant, M has a unique Kähler–Einstein metric* (*with negative Einstein constant*).

Theorem 7.14 was conjectured by Calabi and proved independently by Aubin [Au2] and Yau [Ya2], [Ya3]. The case of positive first Chern class is complicated, we discuss it briefly in Subsection 7.4 below. The case of Ricci-flat Kähler metrics is treated in Corollary 7.2.

We now reformulate Theorem 7.14 into an analytic problem similar to (7.13) above. Suppose $c_1(M) < 0$ and choose a constant $\lambda < 0$. Let ω be a positive real differential form of type $(1,1)$ with $\lambda\omega \in 2\pi c_1(M)$. Then $g(X,Y) = \omega(X,JY)$ is a Kähler metric on M with Kähler form ω.

Let g' be a Kähler–Einstein metric on M with Kähler form ω' and Einstein constant λ. Then $\lambda\omega' = \rho' \in 2\pi c_1(M)$, hence ω' is cohomologous to ω, and therefore $\omega' - \omega = i\partial\bar\partial\varphi$. Since the Ricci form $\rho \in 2\pi c_1(M)$, there is a smooth real function f on M, unique up to an additive constant, such that $\rho - \lambda\omega = i\partial\bar\partial f$. We want to solve $\rho' = \lambda\omega'$. The latter gives

$$-i\partial\bar\partial \ln M(\varphi) = \rho' - \rho = \lambda(\omega' - \omega) - i\partial\bar\partial f = \lambda i\partial\bar\partial\varphi - i\partial\bar\partial f, \tag{7.15}$$

and hence, by Exercise 5.51,

$$\ln M(\varphi) = -\lambda\varphi + f + \text{const}. \tag{7.16}$$

Replacing φ by $\text{const}/\lambda + \varphi$, the constant on the right side vanishes. Vice versa, suppose φ solves (7.16). Then the above computation shows that $\omega' = \omega + i\partial\bar\partial\varphi$ solves $\rho' = \lambda\omega'$. We explain after Equation 7.18 below why $\omega + i\partial\bar\partial\varphi$ is positive. In conclusion, Theorem 7.14 is a consequence of the following assertion:

Given $\lambda < 0$ and a smooth function f on M, there is a unique smooth real function φ on M such that

$$\ln M(\varphi) = -\lambda\varphi + f. \tag{7.17}$$

Both equations, (7.13) and (7.17), are of the form

$$\ln M(\varphi) = f(p,\varphi), \tag{7.18}$$

where $f = f(p,\varphi)$ is a smooth real function [18]. Since $\exp(f(p,\varphi)) > 0$, the same argument as further up shows that for any smooth solution φ of (7.18), $\omega + i\partial\bar\partial\varphi$ is a positive $(1,1)$-form. With this question out of the way, we can now concentrate on the solvability of our equations. If not specified otherwise, inner products and $*$-operator are taken with respect to the metric g.

7.1 Uniqueness. Let $p \in M$. Given a second Kähler metric g' about p, there exists a centered holomorphic chart $z = (z^1, \dots, z^m)$ about p with

$$g_p = \frac{1}{2}\sum dz_p^j \odot d\bar z_p^j, \quad g'_p = \frac{1}{2}\sum a_j\, dz_p^j \odot d\bar z_p^j, \tag{7.19}$$

where $a_1, \dots, a_m$ are positive numbers. Correspondingly,

$$\omega_p = \frac{i}{2}\sum dz_p^j \wedge d\bar z_p^j, \quad \omega'_p = \frac{i}{2}\sum a_j\, dz_p^j \wedge d\bar z_p^l. \tag{7.20}$$

[18] It is understood that $f(p,\varphi) = f(p,\varphi(p))$.

The volume forms are related by

$$\operatorname{vol}'(p) = (\textstyle\prod a_j) \cdot \operatorname{vol}(p). \tag{7.21}$$

We also see that $\omega_p^{m-j-1} \wedge (\omega'_p)^j$ is a positive linear combination of the forms

$$\frac{i^{m-1}}{2^{m-1}} dz_p^1 \wedge d\bar{z}_p^1 \wedge \cdots \wedge \widehat{dz_p^k \wedge d\bar{z}_p^k} \wedge \cdots \wedge dz_p^m \wedge d\bar{z}_p^m, \tag{7.22}$$

where behatted terms are to be deleted.

7.23 Lemma. *With respect to the $*$-operator of g, $*(\omega^{m-j-1} \wedge (\omega')^j)$ is a positive $(1,1)$-form.*

Proof. Immediate from (7.22). □

7.24 Lemma. *Let φ be a smooth real function on M. If η is a differential form of degree $2m-2$, then*

$$d\varphi \wedge d_c\varphi \wedge \eta = *\eta(\operatorname{grad}\varphi, J \operatorname{grad}\varphi) \operatorname{vol}. \tag{1}$$

In particular,

$$d\varphi \wedge d_c\varphi \wedge \omega^{m-1} = \frac{1}{m} |\operatorname{grad}\varphi|^2 \omega^m. \tag{2}$$

Note that if η is of type $(m-1, m-1)$ and $*\eta$ is non-negative (of type $(1,1)$), then $*\eta(\operatorname{grad}\varphi, J\operatorname{grad}\varphi) \geq 0$.

Proof of Lemma 7.24. We have $d_c\varphi = -d\varphi \circ J$. Hence $(d_c\varphi)^\sharp = J \cdot \operatorname{grad}\varphi$, and therefore

$$\begin{aligned} d\varphi \wedge d_c\varphi \wedge \eta &= \langle d\varphi \wedge d_c\varphi, *\eta \rangle \operatorname{vol} \\ &= \langle d_c\varphi, \operatorname{grad}\varphi \llcorner *\eta \rangle \operatorname{vol} \\ &= *\eta(\operatorname{grad}\varphi, J\operatorname{grad}\varphi) \operatorname{vol}. \end{aligned}$$

This proves the first equation. The second equation follows from the first and Equations 4.20 and 4.22. □

We now come to the uniqueness of solutions of (7.18). It is related to the dependence of f on the variable φ.

7.25 Proposition. *Suppose φ_1 and φ_2 solve* (7.18).
1) *If f is weakly monotonically increasing in φ, then $\varphi_1 - \varphi_2$ is constant.*
2) *If f is strictly monotonically increasing in φ, then $\varphi_1 = \varphi_2$.*

In particular, solutions to Equation 7.17 are unique and solutions to Equation 7.13 are unique up to an additive constant.

Proof of Proposition 7.25. Set $\omega_1 = \omega + i\partial\bar\partial\varphi_1$ and $\omega_2 = \omega + i\partial\bar\partial\varphi_2$. Then

$$\int \omega_1^m = \int \omega^m = \int \omega_2^m,$$

where we integrate over all of M. Hence

$$\begin{aligned} 0 &= 2\int_M (\omega_1^m - \omega_2^m) \\ &= \int dd_c(\varphi_1 - \varphi_2) \wedge (\omega_1^{m-1} + \omega_1^{m-2} \wedge \omega_2 + \dots + \omega_2^{m-1}). \end{aligned}$$

Applying Stokes' theorem and Lemmas 7.23 and 7.24 we get

$$\begin{aligned} 0 &= \int d((\varphi_1 - \varphi_2) d_c(\varphi_1 - \varphi_2) \wedge (\omega_1^{m-1} + \omega_1^{m-2} \wedge \omega_2 + \dots + \omega_2^{m-1})) \\ &= \int d(\varphi_1 - \varphi_2) \wedge d_c(\varphi_1 - \varphi_2) \wedge (\omega_1^{m-1} + \omega_1^{m-2} \wedge \omega_2 + \dots + \omega_2^{m-1}) \\ &\quad + \int (\varphi_1 - \varphi_2)(\omega_1^m - \omega_2^m) \\ &\geq \frac{1}{m}\int |\operatorname{grad}(\varphi_1 - \varphi_2)|_{g_1}^2 \omega_1^m + \int (\varphi_1 - \varphi_2)(e^{f(p,\varphi_1)} - e^{f(p,\varphi_2)})\omega^m, \end{aligned}$$

where the index g_1 means that we take the norm with respect to the metric g_1. If f is weakly monotonically increasing in φ, then both terms on the right hand side are non-negative. □

7.2 Regularity. Let $U \subset C^2(M)$ be the open subset of functions φ such that $\omega + i\partial\bar\partial\varphi$ is positive. Let $\mathcal{F}: U \to C^0(M)$ be the functional

$$\mathcal{F}(\varphi) = \ln M(\varphi) - f(p, \varphi), \tag{7.26}$$

where $f = f(p, \varphi)$ is a smooth real function. We want to study the solvability of the equation $\mathcal{F}(\varphi) = 0$ using the continuity method. To that end, the regularity of solutions will play an essential role in Subsection 7.3 since we will need a priori estimates of their higher order derivatives, compare (7.40) and (7.41).

For $\varphi \in U$ and $\omega_\varphi = \omega + i\partial\bar\partial\varphi$, a continuous differential form, we let g_φ be the associated continuous Riemannian metric as in (4.2). Since ω_φ is of type $(1, 1)$, g_φ is compatible with the complex structure of M. We call

$$\Delta_\varphi = -2g_\varphi^{j\bar k}\frac{\partial^2}{\partial z_j \partial \bar z_k}, \tag{7.27}$$

the Laplacian of g_φ, compare Exercise 5.51.1. Since $\varphi \in U$, Δ_φ is elliptic.

For $\varphi \in U$ and $\psi \in C^2(M)$, the directional derivative of $\mathcal{F}$ at $\varphi \in U$ in the direction of ψ is

$$(\mathcal{F}(\varphi + t\psi))'|_{t=0} = -\frac{1}{2}\Delta_\varphi\psi - f_\varphi(p, \varphi)\cdot\psi, \tag{7.28}$$

where f_φ denotes the partial derivative of f in the direction of the variable φ. It follows that $\mathcal{F}$ is continuously differentiable with derivative $d\mathcal{F}(\varphi)\cdot\psi$ given by the right hand side of (7.28). The same statements also hold for $\mathcal{F}$ when considered as a functional from $U \cap C^{k+2+\alpha}(M)$ to $C^{k+\alpha}(M)$, where $k \geq 0$ is an integer and $\alpha \in (0, 1)$.

7.29 Proposition. *If $\varphi \in U \cap C^{2+\alpha}(M)$ solves $\mathcal{F}(\varphi) = 0$, then φ is smooth.*

Proof. Since f is smooth and φ is $C^{2+\alpha}$, the function

$$\det(g_{j\bar{k}})^{-1}\det(g_{j\bar{k}} + \varphi_{j\bar{k}}) - f(p, \varphi)$$

is $C^{2+\alpha}$ in $p \in M$ and the (free) variables $\varphi_{j\bar{k}}$. By (7.28), φ is an elliptic solution of the equation $\mathcal{F}(\varphi) = 0$. Hence φ is $C^{4+\alpha}$ and, recursively, $C^{k+\alpha}$ for any $k \geq 6$, by the regularity theory for elliptic solutions of partial differential equations [19]. □

7.3 Existence. From now on, we only consider Equations 7.13 and 7.17,

$$\mathcal{F}(\varphi) = \ln M(\varphi) + \lambda\varphi - f = 0 \quad \text{on } U \cap C^{2+\alpha}(M), \tag{7.30}$$

where U is as above, $\alpha \in (0, 1)$, f is a given smooth function on M, and $\lambda = 0$ in the case of (7.13) and $\lambda < 0$ in the case of (7.17). We also recall that in the case of $\lambda = 0$, we are only interested in f modulo an additive constant.

7.31 Remark. In the case of complex dimension $m = 1$, Equation 7.30 becomes

$$e^{f-\lambda\varphi}g = g - \frac{1}{2}\Delta\varphi.$$

In this case, the arguments below are much easier. It is a good exercise to follow the argument below in this more elementary case. This is still a non-trivial case, the uniformization of closed Riemann surfaces.

For the proof of the existence of solutions we apply the method of continuity. We let $0 \leq t \leq 1$ and replace f by tf in (7.30). That is, we consider the equation

$$\mathcal{F}_t(\varphi) = \ln M(\varphi) + \lambda\varphi - tf = 0 \quad \text{on } U \cap C^{2+\alpha}(M). \tag{7.30.t}$$

We let $\mathcal{T}_\alpha$ be the set of $t \in [0, 1]$ for which a $C^{2+\alpha}$-solution of Equation 7.30.t exists (modulo a constant function if $\lambda = 0$). The constant function $\varphi = 0$ solves 7.30.0. Hence $\mathcal{T}_\alpha$ is not empty. The aim is now to show that $\mathcal{T}_\alpha$ is open and closed. We distinguish between the cases $\lambda < 0$ and $\lambda = 0$.

We start with the openness and let $t \in \mathcal{T}_\alpha$ and φ be a solution of Equation 7.30.t. Now

$$d\mathcal{F}_t(\varphi) = d\mathcal{F}(\varphi) = -\frac{1}{2}\Delta_\varphi + \lambda \tag{7.32}$$

[19] In [Ka2], Kazdan gives a nice exposition (with references) of this theory.

defines a self-adjoint operator on $L^2(M, g_\varphi)$. If $\lambda < 0$, then the kernel of $d\mathcal{F}_t(\varphi)$ on $L^2(M)$ is trivial and hence $d\mathcal{F}_t(\varphi)$ is an isomorphism from its domain $H^2(M)$ of definition onto $L^2(M)$, by the spectral theorem. Therefore, by the regularity theory for linear partial differential operators, $d\mathcal{F}_t(\varphi)\colon C^{2+\alpha}(M) \to C^\alpha(M)$ is an isomorphism as well. Hence by the implicit function theorem, there is a unique solution φ_s of Equation 7.30.s which is close to φ_t in $C^{2+\alpha}(M)$, for all s close enough to t. Hence $\mathcal{T}_\alpha$ is open.

In the case $\lambda = 0$, we only need to solve modulo constant functions. We let $Z \subset L^2(M)$ be the space of functions with g_φ-mean zero, a closed complement of the constant functions in $L^2(M)$. Since constant functions are smooth, $Z \cap H^2(M)$ and $Z \cap C^{k+\alpha}(M)$ are closed complements of the constant functions in $H^2(M)$ and $C^{k+\alpha}(M)$, for all $k \geq 0$. By the spectral theorem, $d\mathcal{F}_t(\varphi) = -\Delta_\varphi/2$ is surjective from $H^2(M)$ onto Z. Hence $d\mathcal{F}_t(\varphi)\colon C^{2+\alpha}(M) \to Z \cap C^\alpha(M)$ is surjective, by the regularity theory for elliptic differential operators. Hence by the implicit function theorem, there is a family of functions φ_s, for s close enough to t, such that φ_s solves Equation 7.30.s up to constant functions. Hence $\mathcal{T}_\alpha$ is open in this case as well.

We now come to the heart of the matter, the closedness of $\mathcal{T}_\alpha$. We need to derive a priori estimates of solutions φ of Equation 7.30.t, or, what will amount to the same, of Equation 7.30.

7.33 Lemma. *Let $\lambda < 0$ and suppose that* $\ln M(\varphi) = -\lambda\varphi + f$. *Then*

$$\|\varphi\|_\infty \leq \frac{1}{|\lambda|}\|f\|_\infty.$$

Proof. Let $p \in M$ be a point where φ achieves its maximum. Then $M(\varphi)(p) \leq 1$, by (7.12), and hence $\lambda\varphi(p) \geq f(p)$. Since $\lambda < 0$,

$$\varphi \leq \varphi(p) \leq \frac{1}{\lambda} f(p) \leq \frac{1}{|\lambda|}\|f\|_\infty.$$

If φ achieves a minimum at p, then $M(\varphi)(p) \geq 1$, and hence $\lambda\varphi(p) \leq f(p)$. Therefore

$$\varphi \geq \varphi(p) \geq \frac{1}{\lambda} f(p) \geq -\frac{1}{|\lambda|}\|f\|_\infty. \qquad \square$$

The case $\lambda = 0$ is much harder.

7.34 Lemma. *Suppose that φ is a smooth function on M with* $\ln M(\varphi) = f$ *and mean* 0 *with respect to g. Then $\|\varphi\|_\infty \leq C$, where C is a constant which depends only on M, g, and an upper bound for $C(f) := \|1 - \exp\circ f\|_\infty$.*

It will become clear in the proof that the constant C is explicitly computable in terms of $C(f)$ and dimension, volume, and Sobolev constants of M.

Proof of Lemma 7.34. We let φ be a smooth function on M with $\ln M(\varphi) = f$. Recall that the latter holds iff $(\omega')^m = e^f \omega^m$. We have

$$\begin{aligned}(1 - e^f)\omega^m &= \omega^m - (\omega')^m \\ &= (\omega - \omega') \wedge (\omega^{m-1} + \omega^{m-2} \wedge \omega' + \cdots + (\omega')^{m-1}) \\ &= -\frac{1}{2} dd_c\varphi \wedge (\omega^{m-1} + \omega^{m-2} \wedge \omega' + \cdots + (\omega')^{m-1}).\end{aligned}$$

For $p \geq 2$, Stokes' theorem gives

$$\int_M d\{\varphi|\varphi|^{p-2} d_c\varphi \wedge (\omega^{m-1} + \cdots + (\omega')^{m-1})\} = 0,$$

where we note that $\varphi|\varphi|^{p-2}$ is C^1 for $p \geq 2$. Now

$$d(\varphi|\varphi|^{p-2}) = (p-1)|\varphi|^{p-2} d\varphi,$$

hence

$$(p-1)\int_M |\varphi|^{p-2} d\varphi \wedge d_c\varphi \wedge (\omega^{m-1} + \cdots + (\omega')^{m-1}) = 2\int_M (1 - e^f)\varphi|\varphi|^{p-2}\omega^m.$$

By Lemma 7.24, the left hand side is equal to

$$\frac{p-1}{m}\int_M |\varphi|^{p-2}(|\operatorname{grad}\varphi|^2 + \Phi)\omega^m,$$

where $\Phi \geq 0$. From

$$\frac{1}{4}p^2|\varphi|^{p-2}|\operatorname{grad}\varphi|^2 = |\operatorname{grad}|\varphi|^{p/2}|^2$$

and $\omega^m = m!\operatorname{vol}_g$, we conclude that, for $p \geq 2$,

$$\|\operatorname{grad}|\varphi|^{p/2}\|_2^2 \leq \frac{mp^2}{2(p-1)} C(f)\|\varphi\|_{p-1}^{p-1}, \tag{7.35}$$

where the index refers to the corresponding integral norm.

By the Sobolev embedding theorem, there is a constant C_1 with

$$\|\varphi\|_{\frac{2m+2}{m}} \leq C_1 \cdot (\|\varphi\|_2 + \|\operatorname{grad}\varphi\|_2). \tag{7.36}$$

From now on we assume that φ has mean 0 with respect to g. Then we have

$$\|\varphi\|_2 \leq C_2 \cdot \|\operatorname{grad}\varphi\|_2, \tag{7.37}$$

where $1/C_2^2$ is the first eigenvalue of the Laplacian of (M, g).

By (7.35), applied in the case $p = 2$, we have

$$\|\operatorname{grad}\varphi\|_2^2 \leq 2mC(f)\|\varphi\|_1.$$

Now $\|\varphi\|_1 \le \mathrm{vol}(M)^{1/2} \cdot \|\varphi\|_2$. Hence, by (7.37),

$$\|\operatorname{grad}\varphi\|_2^2 \le 2mC(f)C_2 \,\mathrm{vol}(M)^{1/2} \|\operatorname{grad}\varphi\|_2.$$

With $C_3 := 2mC_2 \,\mathrm{vol}(M)^{1/2}$, we get

$$\|\operatorname{grad}\varphi\|_2 \le C_3 C(f) \quad \text{and} \quad \|\varphi\|_2 \le C_2 C_3 C(f),$$

where we apply (7.37) for the second estimate. Hence, by (7.36),

$$\|\varphi\|_{\frac{2m+2}{m}} \le C_1(C_2+1)C_3 C(f). \tag{7.38}$$

We now apply an iteration argument to get the desired uniform bound for φ. With $\alpha = (m+1)/m$ and $p \ge 2\alpha$, we have

$$\begin{aligned}
\|\varphi\|_{p\alpha}^p &= \Big(\int_M |\varphi|^{p\alpha}\Big)^{1/\alpha} = \Big(\int_M |\varphi|^{\frac{p}{2}2\alpha}\Big)^{\frac{2}{2\alpha}} = \||\varphi|^{p/2}\|_{2\alpha}^2 \\
&\le 2C_1^2 \cdot \big(\|\varphi\|_p^p + \|\operatorname{grad}|\varphi|^{p/2}\|_2^2\big) \\
&\le 2C_1^2 \cdot \Big(\|\varphi\|_p^p + \frac{mp^2}{2(p-1)} C(f)\|\varphi\|_{p-1}^{p-1}\Big) \\
&\le 2C_1^2 \cdot \Big(\|\varphi\|_p^p + \frac{mp^2}{2(p-1)} C(f)(\mathrm{vol}\, M)^{1/p}\|\varphi\|_p^{p-1}\Big) \\
&\le C_4 p \cdot \max\{1, \|\varphi\|_p^p\}.
\end{aligned}$$

By Equation 7.38, there is a constant $C \ge 1$ such that

$$\|\varphi\|_{2\alpha} \le C \cdot (C_4 2\alpha)^{-\frac{m+1}{2\alpha}}.$$

We choose C such that

$$C \cdot (C_4 p)^{-\frac{m+1}{p}} \ge 1$$

for all $p \ge 2\alpha$ and assume recursively that

$$\|\varphi\|_p \le C \cdot (C_4 p)^{-\frac{m+1}{p}}$$

for some $p \ge 2\alpha$. Since $m\alpha = m+1$, we then get that

$$\|\varphi\|_{p\alpha}^{p\alpha} \le C_4^\alpha p^\alpha C^{p\alpha}(C_4 p)^{-(m+1)\alpha} = C^{p\alpha}(C_4 p)^{-m-1}$$

and hence that

$$\|\varphi\|_{p\alpha} \le C(C_4 p)^{-\frac{m+1}{p\alpha}}.$$

Since $\|\varphi\|_p \to \|\varphi\|_\infty$ as $p \to \infty$, we conclude that $\|\varphi\|_\infty \le C$. □

For $\lambda \le 0$ fixed, Lemmas 7.33 and 7.34 give an priori estimate on the C^0-norm of solutions φ of Equation 7.30,

$$\|\varphi\|_\infty \le C(\|f\|_\infty, M, g), \tag{7.39}$$

where we assume that the mean of φ is 0 if $\lambda = 0$. This is the main a priori estimate. There are two further a priori estimates. They are less critical[20] and hold, in particular, for positive λ as well. However, their derivation is space and time consuming and will therefore not be presented here. At this stage, the reader should be well prepared to turn to the literature to get the remaining details.

Let $0 < \beta < 1$. The first of the two remaining a priori estimates concerns the g-Laplacian of φ,

$$\|\Delta\varphi\|_\infty \le C'(\|\varphi\|_\infty, \|f\|_\infty, \|\Delta f\|_\infty, |\lambda|, M, g), \tag{7.40}$$

see [Au2], [Ya3] or [Au3, §7.10]. By the definition of the metric g_φ associated to φ, (7.40) gives a uniform upper bound for g_φ against the given metric g. Now the volume elements of g_φ and g only differ by the factor $\exp \circ f$, see (7.5), hence (7.40) gives also a lower bound for g_φ against g. In particular, (7.40) gives estimates between norms associated to g and g_φ.

The second a priori estimate concerns the g-covariant derivative of $dd_c\varphi$,

$$\|\nabla dd_c\varphi\|_\infty \le C''(\|\Delta\varphi\|_\infty, \|f\|_{C^3(M)}, |\lambda|, M, g), \tag{7.41}$$

[Ca3], [Au2], [Ya3] or [Au3, §7.11]. In the derivation of this latter estimate, one actually estimates $\nabla dd_c\varphi$ first in the C^0-norm associated to g_φ. By what we said above, this is equivalent to the estimate in (7.41). The derivation of this estimate is, among others, a true fight against notation and in our sources, the authors refer to the original articles cited above.

From (7.41) and Exercise 5.51.1 we conclude that

$$\|\nabla\Delta\varphi\|_\infty \le C''(\|\Delta\varphi\|_\infty, \|f\|_{C^3(M)}, |\lambda|, M, g). \tag{7.42}$$

Hence $\Delta\varphi$ is uniformly bounded in the C^β-norm associated to the fixed metric g. Using the Schauder estimates, we conclude that φ is bounded in the $C^{2+\beta}$-norm associated to g, where the bound now also depends on β.

Note that in our application of the continuity method, we consider the family of functions tf, $0 \le t \le 1$, in place of the original f, whose norms are bounded by the corresponding norms of f.

We are ready for the final step in the proof of Theorems 7.1 and 7.14. It remains to show that $\mathcal{T}_\alpha$ is closed. Let $0 < \alpha < \beta < 1$. Let $t_n \in \mathcal{T}_\alpha$ and suppose $t_n \to t \in [0, 1]$. Let φ_n be a solution of Equation 7.30.t$_n$. In the case $\lambda = 0$, we assume that φ_n has mean 0. Then by the above, there is a uniform bound on the $C^{2+\beta}$-norm of the functions φ_n. Now the inclusion $C^{2+\beta}(M) \to C^{2+\alpha}(M)$ is compact. Hence by passing to a subsequence if necessary, φ_n converges in $C^{2+\alpha}(M)$. The limit $\varphi = \lim \varphi_n$ solves 7.30.t. Hence $\mathcal{T}_\alpha$ is closed.

[20] An examination of the case $m = 1$ is recommended.

7.4 Obstructions. In Corollary 7.2 and Theorem 7.14, we obtained existence of Kähler–Einstein metrics on closed Kähler manifolds assuming non-positve first Chern class. In the case of positive first Chern class, the arguments break down to a large extent. When applying the continuity method as explained in the previous subsection, it is no longer possible to show openness or closedness of $\mathcal{T}_\alpha$ along the lines of what is said there.

Concerning the openness of $\mathcal{T}_\alpha$, the operator $\Delta_\varphi - 2\lambda$ might have a kernel for $\lambda > 0$. This problem can be overcome [Au4]: In the notation of Subsection 7.3, we need to solve the Monge–Ampère equation $\ln M(\varphi) + \lambda\varphi - f = 0$ with $\lambda > 0$ and a given smooth real function f. In the $\lambda \leq 0$ case, we replaced f with tf, so that we obtained a (trivially) solvable equation for $t = 0$. Here the trick is to keep f, but to replace φ with $t\varphi$. Then the equation for $t = 0$ is highly non-trivial, but solvable according to our previous results for $\lambda = 0$. A simple calculation shows that if ω_t is a solution, then the Ricci form ρ_t of ω_t is given by

$$\rho_t = \lambda t\omega_t + \lambda(1-t)\omega_0,$$

therefore $\mathrm{Ric}_t > \lambda t g_t$ for $t < 1$ and $\mathrm{Ric}_1 = \lambda g_1$. Hence the linearized operator $\Delta_t - 2\lambda t$ is invertible for all $t \in [0, 1)$ for which a solution exists, see Theorem 6.14. This is enough for the continuity method to work, provided we can ensure that $\mathcal{T}_\alpha$ is closed. Observe that if the manifold admits automorphic vector fields, the linearized operator has a kernel for $t = 1$, see Theorem 6.16.

As for the closedness of $\mathcal{T}_\alpha$, the a priori estimates (7.40) and (7.41) do not depend on the sign of λ and continue to hold. However, the remaining C^0-estimate (7.40) is critical and might fail. Indeed, there are closed complex manifolds with positive first Chern class which do not admit Kähler–Einstein metrics at all. As we will see, this can be deduced from special properties of their automorphism groups, a kind of guiding theme: Non-discrete groups of automorphisms yield obstructions to the existence of Kähler–Einstein metrics. Even though a general criterion for the solvability of the Kähler–Einstein problem cannot possibly be based on the automorphism group alone – manifolds with trivial and non-trivial automorphism groups can, but need not be obstructed –, the underlying ideas related to holomorphic actions of complex Lie groups seem to be at the heart of the (still conjectural) full solution.

In the remainder of this subsection, we will treat the two classical obstructions against positive Kähler–Einstein metrics that rely on the automorphism group.

7.43 Theorem (Matsushima [Ma1]). *Let M be a closed Kähler–Einstein manifold. Let $\mathfrak{a}$ be the complex Lie algebra of automorphic vector fields on M and $\mathfrak{k} \subset \mathfrak{a}$ be the real Lie subalgebra of Killing fields. Then $\mathfrak{a} = \mathfrak{k} + J\mathfrak{k}$. If the Einstein constant of M is positive, then $\mathfrak{a} = \mathfrak{k} \oplus J\mathfrak{k}$.*

7.44 Exercise. Verify this explicitly for the Fubini–Study metric on $\mathbb{C}P^m$.

Proof of Theorem 7.43. Let X be a vector field on M and $\xi := X^\flat$. By Hodge theory, $\xi = \eta + \zeta$, where $\eta \in \operatorname{im} d$ and $\zeta \in \ker d^*$. In other words, $Y := \eta^\sharp$ is a gradient field and $Z := \zeta^\sharp$ is volume preserving (recall that $\operatorname{div} Z = -d^*\zeta$).

Assume now that X is automorphic. Our first goal is to prove that Y and Z are automorphic as well. Now Proposition 4.79 and the Bochner identity 1.38 imply that

$$dd^*\eta + d^*d\zeta = \Delta_d\eta + \Delta_d\zeta = \Delta_d\xi = 2\operatorname{Ric}\xi = 2\lambda\xi = 2\lambda\eta + 2\lambda\zeta,$$

where λ is the Einstein constant of M. On the other hand, $\eta \in \operatorname{im} d$ and $\zeta \in \ker d^*$, so $dd^*\eta = 2\lambda\eta$ and $d^*d\zeta = 2\lambda\zeta$ according to the Hodge decomposition theorem, and hence $\Delta\eta = 2\operatorname{Ric}\eta$ and $\Delta\zeta = 2\operatorname{Ric}\zeta$. Again by Proposition 4.79, Y and Z are automorphic.

Now Z is a volume preserving automorphic field, therefore a Killing field, by Theorem 4.81. Hence ∇Y is symmetric and ∇Z is skew-symmetric. Let $\mathfrak{s}$ be the real vector space of all holomorphic vector fields Y such that ∇Y is symmetric. By Exercise 7.45.1, multiplication by J yields an isomorphism $\mathfrak{k} \to \mathfrak{s}$ of real vector spaces. Hence $\mathfrak{a} = \mathfrak{k} + J\mathfrak{k}$.

Now $\mathfrak{k} \cap \mathfrak{s}$ is the space of parallel holomorphic vector fields. If $\lambda > 0$, the Ricci curvature of M is positive, and then M does not carry parallel vector fields. Hence $\mathfrak{a} = \mathfrak{k} \oplus J\mathfrak{k}$ if $\lambda > 0$. □

7.45 Exercise. Let M be a Kähler manifold and X be a real vector field on M.

1) If X is holomorphic, then JX is a Killing field if and only if ∇X is symmetric.
2) If ∇X is symmetric, then JX is volume preserving.
3) Visualize 1) and 2) pictorially if $\dim_{\mathbb{C}} M = 1$.

A Lie algebra is called *reductive* if it is isomorphic to the direct sum of its center and a semi-simple Lie algebra. A connected Lie group is called *reductive* if its Lie algebra is reductive. We note the following consequence of Theorem 7.43.

7.46 Corollary. *If M is a closed Kähler–Einstein manifold, then the component of the identity of the automorphism group of M is reductive.*

Proof. Let $\mathfrak{a}$ and $\mathfrak{k}$ be as above. Since $\mathfrak{k}$ is the Lie algebra of a compact Lie group, $\mathfrak{k} = \mathfrak{z} \oplus [\mathfrak{k}, \mathfrak{k}]$, where $\mathfrak{z}$ is the center of $\mathfrak{k}$. Moreover, the commutator subalgebra $[\mathfrak{k}, \mathfrak{k}]$ of $\mathfrak{k}$ is semi-simple. Now $\mathfrak{k} \cap J\mathfrak{k}$ consists of parallel vector fields, see Exercise 7.45.1, hence $\mathfrak{k} \cap J\mathfrak{k} \subset \mathfrak{z}$. Therefore

$$\mathfrak{a} = \mathfrak{k} + J\mathfrak{k} = (\mathfrak{z} + J\mathfrak{z}) \oplus ([\mathfrak{k}, \mathfrak{k}] + J[\mathfrak{k}, \mathfrak{k}]).$$

Since $\mathfrak{a} = \mathfrak{k} + J\mathfrak{k}$, the center of $\mathfrak{a}$ is $\mathfrak{z} + J\mathfrak{z}$, and $[\mathfrak{k}, \mathfrak{k}] + J[\mathfrak{k}, \mathfrak{k}]$ is semi-simple. □

7.47 Example. Let M be the once blown up $\mathbb{C}P^2$ as in Example 2.43. The Lie algebra $\mathfrak{a}$ of the automorphism group of M consists of the matrices

$$\begin{pmatrix} 0 & * & * \\ 0 & * & * \\ 0 & * & * \end{pmatrix} \in \mathbb{C}^{3\times 3}.$$

The center of $\mathfrak{a}$ is trivial. However, $\mathfrak{a}$ is not semi-simple since the space of matrices in $\mathfrak{a}$ with zero entries in second and third row is an Abelian ideal. Hence $\mathfrak{a}$ is not reductive, and hence M does not carry Kähler–Einstein metrics.

By an analogous argument, the blow up of $\mathbb{C}P^2$ in two points does not carry a Kähler–Einstein metric either; we leave this as an exercise. The blow up of $\mathbb{C}P^2$ in one or two points carries Kähler metrics with $\mathrm{Ric} > 0$, see [Hit], [Ya1]. Moreover, the blow up of $\mathbb{C}P^2$ in one point carries Riemannian metrics with $\mathrm{Ric} > 0$ and sectional curvatures ≥ 0 [Che] and Einstein metrics with $\mathrm{Ric} > 0$ which are conformal to Kähler metrics [PP].

7.48 Remark. For closed Kähler manifolds of constant scalar curvature, we still have $\mathfrak{a} = \mathfrak{k} + J\mathfrak{k}$, see [Li2]. Hence the argument in the proof of Corollary 7.46 applies and shows that the identity components of the automorphism groups of such manifolds are also reductive. In particular, the once and twice blown up $\mathbb{C}P^2$ do not even admit Kähler metrics of constant scalar curvature.

We now discuss a second obstruction to the existence of positive Kähler–Einstein metrics which is also related to the automorphism group. Let M be a closed Kähler manifold with Kähler form ω and Ricci form ρ.

7.49 Theorem (Futaki [Fu]). *Assume that $\omega \in 2\pi c_1(M)$ and write $\rho - \omega = i\partial\bar{\partial}F_\omega$ with $F_\omega \in \mathcal{E}(M,\mathbb{R})$. Then the map*

$$\mathcal{F}_\omega \colon \mathfrak{a} \to \mathbb{R}, \quad \mathcal{F}_\omega(X) := \int_M X(F_\omega)\omega^m,$$

does not depend on the choice of the Kähler form $\omega \in 2\pi c_1(M)$.

We remark that the function F_ω in the definition of the map $\mathcal{F}_\omega$ exists, and is unique up to a constant, since $\rho \in 2\pi c_1(M) \in H^{1,1}(M,\mathbb{R})$, see Exercise 5.51.

The map $\mathcal{F} = \mathcal{F}_\omega$, with $\omega \in 2\pi c_1(M)$ an arbitrary Kähler form, is called the *Futaki invariant* of M. If ω is a Kähler–Einstein metric, $\omega = \rho$, then F_ω is constant and the Futaki invariant $\mathcal{F}_\omega$ vanishes on $\mathfrak{a}$. The Futaki invariant of both the once and the twice blown up $\mathbb{C}P^2$ is non-zero, see [Ti3, Section 3.2]. Hence we conclude again that these two do not admit Kähler–Einstein metrics.

Let $H_1 \to \mathbb{C}P^1$ and $H_2 \to \mathbb{C}P^2$ be the hyperplane bundles. Let M be the total space M of the projective bundle of the product $H_1 \times H_2 \to \mathbb{C}P^1 \times \mathbb{C}P^2$, a closed and complex manifold of complex dimension 4. Futaki [Fu, §3] shows that M is a Fano manifold with reductive automorphism group but non-zero Futaki invariant. Hence the Matsushima criterion for the non-existence of Kähler–Einstein metrics does not apply in this case, whereas the Futaki obstruction does.

Proof of Theorem 7.49. We let ω_t, $-\varepsilon < t < \varepsilon$, be a smooth family of Kähler forms in $2\pi c_1(M)$. The goal is to show that

$$\partial_t\big(\mathcal{F}_{\omega_t}(X)\big) = 0.$$

By Exercise 5.51, $\omega_t - \omega_0 = i\partial\bar{\partial}\Phi_t$ and $\rho_t - \omega_t = i\partial\bar{\partial}F_t$ for families of smooth functions Φ_t and F_t. From the proof of the dd_c-Lemma 5.50 we see that these families of functions can be chosen to be smooth in t. As in (7.15), we compute

$$-i\partial\bar{\partial}\ln M(\Phi_t) = \rho_t - \rho_0 = i\partial\bar{\partial}(F_t + \Phi_t - F_0),$$

hence $-\ln M(\Phi_t) = F_t + \Phi_t - F_0 + \text{constant}$.

From now on, we mostly suppress the parameter t in our notation and indicate differentiation with respect to t by a dot or by ∂_t. With $\varphi = \dot{\Phi}$ and $f = \dot{F}$ we get $\dot{\omega} = i\partial\bar{\partial}\varphi$ and $\dot{\rho} - \dot{\omega} = i\partial\bar{\partial}f$. The displayed equation above gives

$$\Delta\varphi = 2f + 2\varphi,$$

compare Exercise 5.51. By Exercise 7.50, we have

$$2\partial_t\omega_t^m = 2m\dot{\omega} \wedge \omega^{m-1} = 2mi\partial\bar{\partial}\varphi \wedge \omega^{m-1} = -\Delta\varphi \cdot \omega^m.$$

Thus

$$\begin{aligned} 2\partial_t\big(\mathcal{F}_{\omega_t}(X)\big) &= 2\int_M \big\{(Xf)\omega^m + (XF)\partial_t\omega^m\big\} \\ &= \int_M \big\{X(\Delta\varphi - 2\varphi) - (XF)\Delta\varphi\big\}\omega^m. \end{aligned}$$

From now on, we suppress the measure of integration since it is always ω^m. Because X is holomorphic,

$$\begin{aligned} \int_M X(\Delta\varphi) &= \int_M \langle X, \operatorname{grad}\Delta\varphi\rangle \\ &= \int_M \big\{\operatorname{Ric}(X, \operatorname{grad}\varphi) + \langle X, \nabla^*\nabla(\operatorname{grad}\varphi)\rangle\big\} \\ &= \int_M \big\{\operatorname{Ric}(X, \operatorname{grad}\varphi) + \langle \nabla^*\nabla X, \operatorname{grad}\varphi\rangle\big\} \\ &= 2\int_M \operatorname{Ric}(X, \operatorname{grad}\varphi), \end{aligned}$$

by Exercise 6.17 and Proposition 4.79. It follows that

$$\begin{aligned} \int_M X(\Delta\varphi - 2\varphi) &= 2\int_M \big\{\operatorname{Ric}(X, \operatorname{grad}\varphi) - \langle X, \operatorname{grad}\varphi\rangle\big\} \\ &= 2\int_M \big\{\rho(X, J\operatorname{grad}\varphi) - \omega(X, J\operatorname{grad}\varphi)\big\} \\ &= 2\int_M (i\partial\bar{\partial}F)(X, J\operatorname{grad}\varphi) \\ &= \int_M \{\nabla^2 F(\operatorname{grad}\varphi, X) + \nabla^2 F(J\operatorname{grad}\varphi, JX)\}, \end{aligned}$$

because of $\rho - \omega = i\partial\bar{\partial}F$ and Exercise 7.50. Now

$$\operatorname{div}((XF)\operatorname{grad}\varphi) = \langle\operatorname{grad}(XF), \operatorname{grad}\varphi\rangle - (XF)\Delta\varphi,$$

hence

$$\begin{aligned}\int_M (XF)\,\Delta\varphi &= \int_M \langle \operatorname{grad}(XF), \operatorname{grad}\varphi\rangle \\ &= \int_M \bigl\{\nabla^2 F(\operatorname{grad}\varphi, X) + \langle \nabla_{\operatorname{grad}\varphi} X, \operatorname{grad} F\rangle\bigr\}.\end{aligned}$$

Since X is automorphic and J is parallel,

$$\langle \nabla_{\operatorname{grad}\varphi} X, \operatorname{grad} F\rangle = \langle \nabla_{J\operatorname{grad}\varphi} X, J\operatorname{grad} F\rangle = -\langle \nabla_{J\operatorname{grad}\varphi} JX, \operatorname{grad} F\rangle.$$

Adding up, we get

$$\begin{aligned}2\partial_t\bigl(\mathcal{F}_{\omega_t}(X)\bigr) &= \int_M \bigl\{\nabla^2 F(J\operatorname{grad}\varphi, JX) + \langle \nabla_{J\operatorname{grad}\varphi} JX, \operatorname{grad} F\rangle\bigr\} \\ &= \int_M (J\operatorname{grad}\varphi)(JX(F)) = -\int_M (JX(F))\operatorname{div}(J\operatorname{grad}\varphi).\end{aligned}$$

Now $\operatorname{div} J\operatorname{grad}\varphi = 0$, see Exercise 7.45.2, hence $\partial_t\bigl(\mathcal{F}_{\omega_t}(X)\bigr) = 0$. □

7.50 Exercise. Let M be a Kähler manifold and φ be a smooth function on M. Show that

$$2mi\partial\bar\partial\varphi \wedge \omega^{m-1} = -\Delta\varphi\cdot\omega^m, \tag{1}$$

$$2i\partial\bar\partial\varphi(X, JY) = \nabla^2\varphi(X,Y) + \nabla^2\varphi(JX, JY). \tag{2}$$

The *complex Hessian* of φ is defined by $\operatorname{Hess}_c\varphi(X,Y) = i\partial\bar\partial\varphi(X, JY)$. Write down $\operatorname{Hess}\varphi = \nabla^2\varphi$ and $\operatorname{Hess}_c\varphi$ in Kähler normal coordinates.

7.51 Remark. Calabi generalized the Futaki invariant to general closed Kähler manifolds M: Fix a Kähler metric ω on M and write $\rho_\omega = \gamma_\omega + i\partial\bar\partial F_\omega$, where γ_ω is the unique ω-harmonic form in $2\pi c_1(M)$. Then the *Calabi–Futaki invariant*

$$\mathcal{CF}_\omega\colon \mathfrak{a}\to\mathbb{R}, \quad \mathcal{CF}_\omega(X) := \int_M X(F_\omega)\omega^m, \tag{7.52}$$

does not depend on the choice of ω in its Kähler class. It vanishes if the Kähler class of ω contains a Kähler metric with constant scalar curvature. The previously defined Futaki invariant is the special case where $\omega \in 2\pi c_1(M)$ (if the latter is positive definite). Our proof of Theorem 7.49 follows Futaki's original argument. For a discussion of the more general Calabi–Futaki invariant, see for example Chapter 2.I in [Bes], Section 3.1 in [Ti3], or Section 1.6 in [Bo3].

It follows from work of Siu, Tian, and Yau that among the Fano surfaces as in Example 7.4.2, only the blow ups of $\mathbb{C}P^2$ in one or two points do not admit Kähler–Einstein metrics [Si2], [Ti1], [Ti2], [TY]. These are precisely the ones whose automorphism group is not reductive and whose Futaki invariants do not vanish. In higher dimensions, the situation is much more complicated. There are examples, where the C^0-estimates in the continuity method could be verified. For example, the Fermat hypersurfaces

$$z_0^d + \cdots + z_n^d = 0$$

in $\mathbb{C}P^n$ carry Kähler–Einstein metrics if $(n+1)/2 \le d \le n$, see [Na], [Si2], [Ti1] and compare Example 7.4.1. For surveys on Kähler–Einstein metrics, we refer to [Bo3] and [Ti3]. More recent references are [Bi] and [Th]. All these references contain extensive bibliographies as well.

8 Kähler Hyperbolic Spaces

A question attributed to Chern asks whether the Euler characteristic of a closed Riemannian manifold M of dimension $2m$ satisfies

$$(-1)^m \chi(M) > 0 \tag{8.1}$$

if the sectional curvature K of M is negative. The answer is yes if $m \le 2$, see [Chr] for the case $m = 2$. Dodziuk and Singer remarked that Atiyah's L^2-index theorem implies (8.1) if the space of square integrable harmonic forms on the universal covering space of M vanishes in degree $\neq m$ and does not vanish in degree $= m$. In [Gr], Gromov proves this and more in the case where M is a Kähler manifold; in this section we are concerned with his arguments and results.

Let M be a complete and connected Riemannian manifold and E be a flat Hermitian vector bundle over M. Choose an origin $o \in M$ and let $r = d(o, \cdot)$ be the distance to o. Let α be a differential form on M with values in E and $\mu\colon \mathbb{R}_+ \to \mathbb{R}_+$ be a non-decreasing function. We say that α is $O(\mu(r))$ if there is a constant c such that

$$|\alpha| \le c\mu(cr + c) + c, \tag{8.2}$$

where we consider the norm $|\alpha| = |\alpha(p)|$ as a function on M. Following Gromov, we say that α is $d(O(\mu(r)))$ if there is a differential form β on M with $\alpha = d\beta$ and such that β is $O(\mu(r))$. We also write d(bounded) instead of $d(O(1))$, that is, in the case where β can be chosen to be uniformly bounded.

8.3 Examples. 1) Let $M = \mathbb{R}^n$ with Euclidean metric and origin 0. Let α be the volume form of M, $\alpha = dx^1 \wedge \dots \wedge dx^n$. Let β be an $(n-1)$-form with $d\beta = \alpha$. Then on $B = B(r, 0)$,

$$\operatorname{vol}_n B = \int_B \alpha = \int_{\partial B} \beta \le \max\{|\beta(p)| \mid p \in \partial B\} \operatorname{vol}_{n-1} \partial B.$$

It follows that

$$\max\{|\beta(p)| \mid p \in \partial B\} \ge \frac{\operatorname{vol}_n B}{\operatorname{vol}_{n-1} \partial B} = \frac{r}{n}.$$

Hence α is not d(bounded). In a similar way one can treat each of the differential forms dx^I, where I is a multi-index.

2) Let α be a bounded 1-form on M and f be a smooth function with $df = \alpha$. Then

$$|df| = |\alpha| \le \|\alpha\|_\infty := \sup\{|\alpha(p)| \mid p \in M\}.$$

Hence f is Lipschitz continuous with Lipschitz constant $\|\alpha\|_\infty$ if $\|\alpha\|_\infty < \infty$.

8.4 Proposition. *Let M be complete and simply connected and α be a differential form on M of degree k with $\|\alpha\|_\infty < \infty$ and $d\alpha = 0$.*

1) *If $K \le 0$, then $\alpha = d(O(r))$.*

2) *If $K \le -c^2 < 0$ and $k \ge 2$, then $\alpha = d(bounded)$.*

Proof. Since M is complete and simply connected with non-positive sectional curvature, the theorem of Hadamard–Cartan implies that $\exp_o\colon T_oM \to M$ is a diffeomorphism. Hence we may define

$$\tau_t(p) := \exp_o(t \exp_o^{-1}(p)),$$

where $0 \le t \le 1$. Comparison with Euclidean space shows that the differential of τ_t satisfies

$$|\tau_{t*}v| \le t|v|.$$

The usual proof of the Poincaré lemma provides a $(k-1)$-form β with $d\beta = \alpha$. In Riemannian normal coordinates, β is given by

$$\beta(x) = r \int_0^1 t^{k-1}(\nu \llcorner \alpha)(tx)\, dt = r \int_0^1 \tau_t^*(\nu \llcorner \alpha)(x)\, dt,$$

where $r = |x|$ and ν is the radial normal field. By the above estimate

$$|\tau_t^*(\nu \llcorner \alpha)| \le t^{k-1}|\nu \llcorner \alpha| \le t^{k-1}\|\alpha\|_\infty,$$

and hence

$$|\beta(p)| \le \frac{r}{k}\|\alpha\|_\infty,$$

where $p = \exp(x)$. Note that this estimate is optimal, see Example 8.3.1 above.

Suppose now that the sectional curvature of M is negative, $K \le -c^2 < 0$. Comparison with hyperbolic space shows that, for v perpendicular to ν,

$$|\tau_{t*}v| \le \frac{\sinh(ctr)}{\sinh(cr)}|v|,$$

where r is the distance of the foot point of v to o. Hence

$$\begin{aligned}
|\beta(p)| &\le r \int_0^1 \Big(\frac{\sinh(ctr)}{\sinh(cr)}\Big)^{k-1} dt \cdot \|\alpha\|_\infty \\
&= \int_0^r \Big(\frac{\sinh(cs)}{\sinh(cr)}\Big)^{k-1} ds \cdot \|\alpha\|_\infty \\
&\le \int_{-\infty}^0 (e^{(k-1)cs})\, ds \cdot \|\alpha\|_\infty = \frac{1}{(k-1)c}\|\alpha\|_\infty,
\end{aligned}$$

therefore $\alpha = d(\text{bounded})$. □

8.5 Remark. The estimates in the proof are explicit. Note also that the lift of a differential form on a compact manifold to any covering manifold is uniformly bounded.

8.6 Proposition. *Let (G, K) be a Riemannian symmetric pair such that $M = G/K$ is a symmetric space of non-compact type. If α is a G-invariant differential form on M, then α is parallel and hence closed. If α is of positive degree, then the restriction of α to maximal flats in M vanishes and $\alpha = d(\text{bounded})$.*

This result seems to be well known to experts, I learned it from Anna Wienhard. The proof (and even the statement) of Proposition 8.6 requires more on symmetric spaces than we develop in Appendix B. Therefore we only give a brief sketch of the argument. Good references for symmetric spaces of non-compact type are Chapter 2 in [Eb] and Chapter VI in [Hel].

Sketch of proof of Proposition 8.6. We can assume that (G, K) is an effective pair. By Remark B.36, G contains the connected component of the identity of the isometry group of M. By Theorem B.24.3, α is parallel and hence closed.

Let $F \subset M$ be a maximal flat and $p \in F$. There is a basis of T_pF such that the reflections of T_pF about the hyperplanes perpendicular to the vectors of the basis are realized by the differentials of isometries $g \in G$ fixing p. Since α is invariant under G, it follows easily that $\alpha | T_pF = 0$.

Let $c\colon \mathbb{R} \to M$ be a regular unit speed geodesic and $b\colon M \to \mathbb{R}$ be the Busemann function centered at $c(\infty)$ with $b(c(t)) = -t$. Let $V = -\operatorname{grad} b$ and (f_t) be the flow of V. Then, for each $p \in M$, $c_p(t) = f_t(p)$, $t \in \mathbb{R}$, is the unit speed geodesic through p asymptotic to c. In particular, c_p is also regular and, therefore, contained in a unique maximal flat F_p. The flats F_p constitute a smooth foliation $\mathcal{F}$ of M by totally geodesic Euclidean spaces.

For any unit tangent vector u of M perpendicular to $\mathcal{F}$, $\langle R(u, V)V, u\rangle$ is negative and, by homogeneity, bounded away from 0 by a negative constant which does not depend on u. In particular, the Jacobi fields of geodesic variations by geodesics c_p and perpendicular to $\mathcal{F}$ decay uniformly exponentially. Since α vanishes along $\mathcal{F}$, it follows easily that

$$\gamma = -\int_0^\infty F_t^* \alpha \, dt$$

is a well defined smooth differential form on M such that $L_V\gamma = \alpha$ and $d\gamma = 0$. With $\beta = V \llcorner \gamma$ we get

$$\alpha = L_V\gamma = d(V \llcorner \gamma) + V \llcorner d\gamma = d\beta.$$

Now γ is uniformly bounded and $|V| = 1$, hence $\alpha = d(\text{bounded})$. □

8.7 Definition (Gromov). Let M be a Kähler manifold with Kähler form ω. We say that M is *Kähler-hyperbolic* if the Kähler form $\tilde{\omega}$ of the universal covering space $\tilde{M}$ of M is d(bounded).

Recall that the cohomology class of the Kähler form of a closed Kähler manifold M is non-trivial. Hence closed Kähler manifolds with finite fundamental group are not Kähler-hyperbolic. On the other hand, if the sectional curvature of M is strictly negative, then M is Kähler-hyperbolic, by Proposition 8.4. By Proposition 8.6, Hermitian symmetric spaces of non-compact type are Kähler hyperbolic as well. These do not have strictly negative curvature if their rank, that is, the dimension of their maximal flats, is larger than 1.

Kähler-hyperbolic Kähler manifolds are the main topic in Gromov's article [Gr]. Cao and Xavier [CX] and independently Jost and Zuo [JZ] observed that the arguments of Gromov concerned with the vanishing of square integrable harmonic forms work also under the weaker assumption that the Kähler form is $d(O(\mu))$, where $\sum_1^\infty 1/\mu(n) = \infty$. Note that this assumption is fulfilled if M is simply connected with non-positive sectional curvature, by Proposition 8.4.1. We include this extension into our discussion, see Theorem 8.9 below.

For the rest of this section we assume that M is a complete and connected Kähler manifold with Kähler form ω and complex dimension m. We let $E \to M$ be a flat Hermitian vector bundle. Declaring parallel sections as holomorphic turns E into a holomorphic vector bundle such that the Chern connection of the given Hermitian metric is the given flat connection.

Without further notice we use notation and results from Appendix C. We do not assume throughout that differential forms are smooth and indicate their regularity when appropriate.

8.1 Kähler Hyperbolicity and Spectrum. One of the remarkable properties of Kähler manifolds is that the Lefschetz map commutes with the Laplacian: If α is a harmonic form with values in E, then $\omega \wedge \alpha$ is harmonic as well. This is one of the cornerstones in the arguments which follow.

8.8 Lemma. *Suppose that ω is $d(O(\mu(r)))$, where $\sum_1^\infty 1/\mu(n) = \infty$. Let α be a closed and square integrable differential form with values in E. Then*

$$d^*(\omega \wedge \alpha) = 0 \implies \omega \wedge \alpha = 0.$$

Proof. Choose an origin $o \in M$, and let $r\colon M \to \mathbb{R}$ be the distance to o. Since M is complete and connected, the sublevels of r are relatively compact and exhaust M. Choose cut off functions $\varphi_n\colon M \to \mathbb{R}$, $n \geq 1$, with $\varphi_n(p) = 1$ if $r(p) \leq n$, $\varphi_n(p) = 0$ if $r(p) \geq n+1$ and $|d\varphi_n| \leq c_0$.

Let $\omega = d\eta$ with $|\eta| = O(\mu(r))$. Since $d\alpha = 0 = d^*(\omega \wedge \alpha)$ and $\varphi_n\eta$ is smooth with compact support,

$$\begin{aligned} 0 &= \int_M (d^*(\omega \wedge \alpha), \varphi_n\eta \wedge \alpha) = \int_M (\omega \wedge \alpha, d(\varphi_n\eta \wedge \alpha)) \\ &= \int_{A_n} (\omega \wedge \alpha, d\varphi_n \wedge \eta \wedge \alpha) + \int_M (\omega \wedge \alpha, \varphi_n\omega \wedge \alpha), \end{aligned}$$

where $A_n := \{p \in M \mid n \leq r(p) \leq n+1\}$ contains the support of $d\varphi_n$. The L^2-norm of $\omega \wedge \alpha$ is finite since ω is parallel and, therefore, uniformly bounded. Hence the second term on the right tends to the square of the L^2-norm of $\omega \wedge \alpha$ as $n \to \infty$. Now

$$\int_{A_n} \big|(\omega \wedge \alpha, d\varphi_n \wedge \eta \wedge \alpha)\big| \leq (c\mu(cn+c)+c) \cdot \|\omega\|_\infty \int_{A_n} |\alpha|^2.$$

Hence the assertion follows if the limes inferior of the sequence on the right hand side is 0. If the latter would not hold, then there would be an $\varepsilon > 0$ such that

$$\|\alpha\|_2^2 = \int_M |\alpha|^2 = \sum_n \int_{A_n} |\alpha|^2 \geq \sum \frac{\varepsilon}{(c\mu(cn+c)+c) \cdot \|\omega\|_\infty} = \infty.$$

This would be in contradiction to the assumption that α is square integrable. □

8.9 Vanishing Theorem (Cao–Xavier [CX], Jost–Zuo [JZ]). *If ω is $d(O(\mu))$, where $\sum_1^\infty 1/\mu(n) = \infty$, then*

$$\mathcal{H}_2^k(M,E) = 0 \quad \text{for } k \neq \dim_{\mathbb{C}} M.$$

Proof. For $k < m$ this follows from Theorem C.49 and Lemma 8.8. For $k > m$, vanishing follows from Poincaré duality. □

8.10 Main Lemma. *Suppose $\omega = d\eta$ with $\|\eta\|_\infty < \infty$. Let $\alpha \in L^2(A^k(M,E))$ be in the domain of $\Delta_{d,\max}$ and suppose that $k \neq m$ or that $\alpha \perp \mathcal{H}_2^m(M,E)$. Then*

$$(\Delta_{d,\max}\alpha, \alpha)_2 \geq \lambda^2 \|\alpha\|_2^2 \quad \text{with } \lambda^2 = c(m)/\|\eta\|_\infty^2 > 0.$$

Proof. Assume first that $k = m + s > m$. The s-th power of the Lefschetz map defines a parallel field of isomorphisms $A^{m-s}(M,E) \to A^{m+s}(M,E)$, see the discussion in Section 5. In particular, there is $\beta \in L^2(A^{m-s}(M,E))$ such that $\alpha = \omega^s \wedge \beta$.

Since E is flat, $\Delta_d = (d+d^*)^2$, hence $\Delta_{d,\max} = \Delta_{d,\min}$, by Theorem C.19. Hence we can assume that α and β are smooth with compact support. Let

$$\theta := \eta \wedge \omega^{s-1} \wedge \beta.$$

Then

$$d\theta = \omega^s \wedge \beta - \eta \wedge \omega^{s-1} \wedge d\beta =: \alpha - \alpha'.$$

Now $|\theta| \leq |\eta||\omega^{s-1}||\beta|$, hence

$$\|\theta\|_2 \leq \|\eta\|_\infty |\omega^{s-1}| \|\beta\|_2,$$

where we note that $|\omega^{s-1}|$ is constant. Since $L^{\pm s}$ is parallel, there is a positive constant c_0 such that $|L^s\gamma| \leq c_0|\gamma|$ for all $\gamma \in A^{m-s}(M,E)$ and $|L^{-s}\gamma| \leq c_0|\gamma|$ for all $\gamma \in A^{m+s}(M,E)$. Since L^s commutes with Δ_d, we have $L^s\Delta_d\beta = \Delta_d\alpha$, hence

$$\|d\beta\|_2^2 \leq (\Delta_d\beta, \beta)_2 \leq \|\Delta_d\beta\|_2\|\beta\|_2 \leq c_0^2\|\Delta_d\alpha\|_2\|\alpha\|_2.$$

We conclude that

$$\|\alpha'\|_2 \leq c_0\|\eta\|_\infty|\omega^{s-1}|\|\Delta_d\alpha\|_2^{1/2}\|\alpha\|_2^{1/2}.$$

We also have

$$\begin{aligned}(d^*\alpha, \theta)_2 &\leq \|d^*\alpha\|_2\|\theta\|_2 \leq (\Delta_d\alpha, \alpha)_2^{1/2}\|\theta\|_2 \\ &\leq \|\eta\|_\infty|\omega^{s-1}|\|\Delta_d\alpha\|_2^{1/2}\|\alpha\|_2^{1/2}\|\beta\|_2 \\ &\leq c_0\|\eta\|_\infty|\omega^{s-1}|\|\Delta_d\alpha\|_2^{1/2}\|\alpha\|_2^{3/2}.\end{aligned}$$

Therefore

$$\begin{aligned}\|\alpha\|_2^2 &= (\alpha, d\theta + \alpha')_2 \\ &= (\alpha, d\theta)_2 + (\alpha, \alpha')_2 \\ &\le (d^*\alpha, \theta)_2 + \|\alpha\|_2\|\alpha'\|_2 \\ &\le 2c_0\|\eta\|_\infty|\omega^{s-1}|\|\Delta_d\alpha\|_2^{1/2}\|\alpha\|_2^{3/2}.\end{aligned}$$

Since $\Delta_{d,\max}$ is self-adjoint and non-negative, this proves the asserted inequality in the case $k > m$. By applying Poincaré duality we conclude that it also holds in the case $k < m$.

We note that for a differential form α of pure degree, we have

$$\alpha \in \operatorname{dom}(d + d^*)_{\max} \quad \text{iff} \quad \alpha \in \operatorname{dom} d_{\max} \cap \operatorname{dom} d^*_{\max}.$$

Since $(d + d^*)_{\max} = (d + d^*)_{\min}$, the above inequality shows that, for $k \ne m$,

$$\|d_{\max}\alpha\|_2^2 + \|d^*_{\max}\alpha\|_2^2 \ge \lambda^2\|\alpha\|_2^2$$

for any differential k-form $\alpha \in \operatorname{dom}(d + d^*)_{\max}$. Now $\overline{\operatorname{im} d} \subset \operatorname{dom} d_{\max}$, hence

$$\begin{aligned}\operatorname{dom} d_{\max} &= \operatorname{dom} d_{\max} \cap (\ker d^*_{\max} + \overline{\operatorname{im} d}) \\ &= \operatorname{dom} d_{\max} \cap \ker d^*_{\max} + \overline{\operatorname{im} d}.\end{aligned}$$

Since $\overline{\operatorname{im} d} \subset \ker d_{\max}$, it follows that the image of $d_{\max}$ on differential forms of degree $m - 1$ is equal to the image of its restriction to

$$\operatorname{dom} d_{\max} \cap \ker d^*_{\max} \cap L^2(A^{m-1}(M, E)).$$

By the above estimate we have

$$\|d_{\max}\alpha\|_2^2 \ge \lambda^2\|\alpha\|_2^2$$

for any α in the latter space. In particular, the image of $d_{\max}$ on differential forms of degree $m - 1$ is closed. Therefore it contains $\overline{\operatorname{im} d} \cap L^2(A^m(M, E))$.

There is a similar discussion for $d^*_{\max}$ on forms of degree $m + 1$. Hence we can represent any $\gamma \in \operatorname{dom} \Delta_{d,\max} \cap L^2(A^m(M, E))$ perpendicular to $\mathcal{H}_2^m(M, E)$ as

$$\gamma = d_{\max}\alpha + d^*_{\max}\beta,$$

where $d_{\max}\alpha \in \overline{\operatorname{im} d} \perp \overline{\operatorname{im} d^*} \ni d^*_{\max}\beta$ and

$$d^*_{\max}\alpha = d_{\max}\beta = 0, \quad \|d_{\max}\alpha\|_2 \ge \lambda^2\|\alpha\|_2, \quad \|d^*_{\max}\beta\|_2 \ge \lambda^2\|\beta\|_2.$$

We have

$$\gamma \in \operatorname{dom} \Delta_{d,\max} = \operatorname{dom} \Delta_{d,\min} \subset \operatorname{dom}(d + d^*)_{\min} = \operatorname{dom}(d + d^*)_{\max}.$$

Considering degrees, we get

$$d_{\max}\alpha \in \operatorname{dom} d^*_{\max}, \quad d^*_{\max}\beta \in \operatorname{dom} d_{\max},$$

and hence $\alpha, \beta \in \operatorname{dom} \Delta_{d,\max}$. In conclusion,

$$\begin{aligned}\|\gamma\|_2^2 &= \|d_{\max}\alpha\|_2^2 + \|d^*_{\max}\beta\|_2^2 \\ &= (\Delta_{d,\max}\alpha, \alpha)_2 + (\Delta_{d,\max}\beta, \beta)_2 \\ &\le \|\Delta_{d,\max}\alpha\|_2\|\alpha\|_2 + \|\Delta_{d,\max}\beta\|_2\|\beta\|_2 \\ &\le \lambda^{-2} \cdot (\|\Delta_{d,\max}\alpha\|_2^2 + \|\Delta_{d,\max}\beta\|_2^2) \\ &= \lambda^{-2} \cdot (\|d^*_{\max}\gamma\|_2^2 + \|d_{\max}\gamma\|_2^2) = \lambda^{-2} \cdot (\Delta\gamma, \gamma)_2. \end{aligned}$$

□

Let $E', E'' \to M$ be Hermitian vector bundles over M and D be a differential operator of first order from $\mathcal{E}(E')$ to $\mathcal{E}(E'')$. Let $F \to M$ be a further Hermitian vector bundle with Hermitian connection ∇[21]. Define

$$S^F : E' \otimes T^*M \otimes F \to E'' \otimes F, \quad S^F(\alpha \otimes d\varphi \otimes \sigma) = (S(d\varphi)\alpha) \otimes \sigma, \tag{8.11}$$

where S is the principal symbol of D, see (C.10). The *twist* of D by ∇ is the differential operator $D^\nabla : \mathcal{E}(E' \otimes F) \to \mathcal{E}(E'' \otimes F)$ defined by

$$D^\nabla(\alpha \otimes \sigma) = (D\alpha) \otimes \sigma + S^F(\alpha \otimes \nabla\sigma). \tag{8.12}$$

8.13 Example. For $D = d + d^*$, we have

$$S^F(\alpha \otimes \nabla\sigma) := \sum\{(X_j^\flat \wedge \alpha) \otimes \nabla_{X_j}\sigma - (X_j \llcorner \alpha) \otimes \nabla_{X_j}\sigma\},$$

where (X_j) is a local orthonormal frame of TM.

If ∇, ∇^0 are Hermitian connections on F and $A = \nabla - \nabla^0$, then

$$(D^\nabla - D^{\nabla^0})(\alpha \otimes \sigma) = S^F(\alpha \otimes A\sigma) =: S^A(\alpha \otimes \sigma) \tag{8.14}$$

is of order 0, a potential perturbing D^{∇^0}. We apply this in the case where $F = M \times \mathbb{C}$ is the trivial line bundle over M with the canonical Hermitian metric and canonical flat and Hermitian connection ∇^0. Let ζ be a real valued differential one-form on M. Then

$$\nabla = \nabla^0 + A, \tag{8.15}$$

where

$$A\sigma := i\zeta \otimes \sigma, \tag{8.16}$$

is a connection form for F. Since $i\zeta$ is purely imaginary, ∇ is Hermitian. The curvature form of ∇ is $id\zeta$. Vice versa, if ∇ is a Hermitian connection on a Hermitian line bundle $F \to M$ and the curvature form of ∇ is exact, say equal to $id\zeta$, then $\nabla^0 = \nabla - i\zeta \otimes \mathrm{id}$ is a flat Hermitian connection (and F is trivial if M is simply connected).

[21] We are short of symbols since D is in use already.

8.17 Lemma. *Suppose $\omega = d\eta$ with $\|\eta\|_\infty < \infty$. Let $D = \bar\partial + \bar\partial^*$ on $\mathcal{A}^*(M, E)$ and λ be as in Lemma* 8.10. *Let $F = M \times \mathbb{C}$ be the trivial line bundle over M with ∇^0 and $\nabla = \nabla^0 + i\zeta \otimes \mathrm{id}$ as above. Assume that $|S^A\beta| \le c \cdot |\beta|$ for some positive constant $c < \lambda/\sqrt{2}$. Then*

$$\ker D_{\max} = 0 \implies \ker D^\nabla_{\max} = 0.$$

Proof. Assume that $\ker D_{\max} = 0$. Since ∇^0 is flat, we have

$$\|D^{\nabla^0}\beta\|_2^2 = (\Delta_{\bar\partial}\beta, \beta)_2 = \frac{1}{2}(\Delta_d\beta, \beta)_2 \ge \frac{\lambda^2}{2}\|\beta\|_2^2$$

for all $\beta \in \mathcal{A}^*_c(M, E \otimes F)$, by Lemma 8.10. Hence, for all such β,

$$\begin{aligned}\|D^\nabla\beta\|_2 &= \|D^{\nabla^0}\beta + S^A\beta\|_2 \\ &\ge \|D^{\nabla^0}\beta\|_2 - \|S^A\beta\|_2 \ge (\lambda/\sqrt{2} - c)\|\beta\|_2.\end{aligned}$$

Now $D^\nabla_{\min} = D^\nabla_{\max}$, see Theorem C.19. Hence $\ker D^\nabla_{\max} = 0$ as asserted. □

8.2 Non-Vanishing of Cohomology. We start with a little detour (and a corresponding change in notation) and discuss Atiyah's L^2-index theorem [At]. Let $\tilde M \to M$ be a normal Riemannian covering, where M and $\tilde M$ are connected and M is closed. Let $\tilde E \to \tilde M$ be the pull back of a Hermitian vector bundle $E \to M$. Then the group Γ of covering transformations of $\tilde M \to M$ acts on $\tilde E$, hence on sections of $\tilde E$ by $\gamma \cdot \sigma := \gamma \circ \sigma \circ \gamma^{-1}$.

Let $H \subset L^2(\tilde M, \tilde E)$ be a closed subspace and (Φ_n) be an orthonormal basis of H. Then the function

$$\tilde f\colon \tilde M \to [0, \infty], \quad \tilde f(p) = \sum |\Phi_n(p)|^2, \tag{8.18}$$

does not depend on the choice of (Φ_n). It follows that $\tilde f$ is Γ-invariant if H is, and then $\tilde f$ is the lift of a function f on M and we set

$$\dim_\Gamma H = \int_M f, \tag{8.19}$$

the Γ-*dimension* of H. It is important that $\dim_\Gamma H \ne 0$ iff $H \ne 0$.

Let $E^\pm \to M$ be Hermitian vector bundles and $D\colon \mathcal{E}(E^+) \to \mathcal{E}(E^-)$ be a differential operator. Let $\tilde E^\pm$ and $\tilde D$ be the pull backs of $E^\pm$ and D to $\tilde M$. Then the kernels of $\tilde D$ and its formal adjoint $\tilde D^*$ are Γ-invariant. By definition, the Γ-index of $\tilde D$ is

$$\mathrm{ind}_\Gamma \tilde D = \dim_\Gamma \ker \tilde D - \dim_\Gamma \ker \tilde D^*. \tag{8.20}$$

The Γ-index of $\tilde D$ is well-defined if one of the dimensions on the right is finite.

8.21 Theorem (Atiyah [At]). *If D is elliptic, then the Γ-dimensions on the right hand side of* (8.20) *are finite and*

$$\mathrm{ind}_\Gamma \tilde D = \mathrm{ind}\, D.$$

Theorem 8.21 applies in the case of non-positively curved closed Kähler manifolds and shows that their Euler characteristic has the right sign. More precisely, we have the following result.

8.22 Theorem (Cao–Xavier [CX], Jost–Zuo [JZ]). *Let M be a closed Kähler manifold and $\tilde{M} \to M$ be its universal cover. If the Kähler form $\tilde{\omega}$ of $\tilde{M}$ is $d(O(\mu))$, where $\sum_1^\infty 1/\mu(n) = \infty$, then*

$$(-1)^m \chi(M, E) \geq 0.$$

Proof. By Hodge theory, $\chi(M, E) = \operatorname{ind} D$, where $D = d + d^*$ from $\mathcal{A}^{\text{even}}(M, E)$ to $\mathcal{A}^{\text{odd}}(M, E)$. We have $\tilde{D} = d + d^*$, but now from $\mathcal{A}^{\text{even}}(\tilde{M}, \tilde{E})$ to $\mathcal{A}^{\text{odd}}(\tilde{M}, \tilde{E})$. By Theorem 8.21, $\operatorname{ind} D = \operatorname{ind}_\Gamma \tilde{D}$, where Γ is the fundamental group of M. By Theorem 8.9, there are no non-trivial square integrable harmonic forms on $\tilde{M}$ of degree different from the middle dimension. Hence

$$(-1)^m \operatorname{ind}_\Gamma \tilde{D} = \dim_\Gamma \mathcal{H}^m(M, E) \geq 0. \qquad \square$$

Gromov uses an extended version of (part of) Theorem 8.21. He considers the twist of the pull back of a specific bundle E with the trivial complex line bundle endowed with a Hermitian connection ∇ which is not invariant under Γ, whereas the curvature of ∇ still is.

For the reader's convenience, but without proof, we state a more general version of Gromov's extension. Let $\tilde{E} \to \tilde{M}$ be a Hermitian vector bundle of rank r, not necessarily the pull back of a Hermitian vector bundle over M. Let G be an extension of Γ by a compact group K,

$$1 \longrightarrow K \longrightarrow G \xrightarrow{\rho} \Gamma \longrightarrow 1. \tag{8.23}$$

Suppose that G acts by Hermitian isomorphisms on $\tilde{E}$ such that

$$\pi \circ g = \rho(g) \circ \pi \quad \text{for all } g \in G, \tag{8.24}$$

where π denotes the projection of $\tilde{E}$. Then G acts on $L^2(\tilde{M}, \tilde{E})$ by

$$g \cdot \sigma := g \circ \sigma \circ \rho(g)^{-1}. \tag{8.25}$$

Let H be a G-invariant subspace of $L^2(\tilde{M}, \tilde{E})$. Then the function $\tilde{f}$ as in (8.18) is Γ-invariant, and we define the G-dimension of H by the integral of the corresponding function f on M as in (8.19).

Let $\tilde{E}^\pm$ be Hermitian vector bundles with Hermitian G-actions satisfying (8.24) (where now π denotes the projections of $\tilde{E}^\pm$). Suppose that $\tilde{D}$ is a differential operator from $\mathcal{E}(\tilde{E}^+)$ to $\mathcal{E}(\tilde{E}^-)$ commuting with the actions of G,

$$\tilde{D}(g \cdot \sigma) = g \cdot (\tilde{D}\sigma). \tag{8.26}$$

Then the kernels of $\tilde{D}$ and $\tilde{D}^*$ are invariant under G, and we define the G-index of $\tilde{D}$ as in (8.20). The case considered in Theorem 8.21 corresponds to the trivial extension

$G = \Gamma$. The extension we state concerns the case where $\tilde{E} = \tilde{E}^+ \oplus \tilde{E}^-$ is a graded Dirac bundle and $\tilde{D}^+ : \mathcal{E}(\tilde{E}^+) \to \mathcal{E}(\tilde{E}^-)$ the associated Dirac operator in the sense of Gromov and Lawson, see Subsection C.24.

8.27 Theorem. *Suppose that $\tilde{E}$ is a graded Dirac bundle and that the Hermitian action of G on $\tilde{E}$ satisfies* (8.24) *and leaves invariant connection, Clifford multiplication, and splitting $\tilde{E} = \tilde{E}^+ \oplus \tilde{E}^-$ of $\tilde{E}$. Then the associated Dirac operator $\tilde{D}$ satisfies* (8.26), *the G-dimensions of* $\ker \tilde{D}^+$ *and* $\ker \tilde{D}^-$ *are finite, the canonical index form $\tilde{\alpha}$ of $\tilde{D}^+$ is the pull back of a form α on M, and*

$$\operatorname{ind}_G(\tilde{D}^+) = \int_M \alpha.$$

About the proof. It seems that the arguments in Section 13 of [Ro] generalize to the setting of Theorem 8.27. □

8.28 Remark. We will need Theorem 8.27 for the Hirzebruch–Riemann–Roch formula, that is, for the twisted Dolbeault operator $\bar{\partial} + \bar{\partial}^*$ (times $\sqrt{2}$) on the Dirac bundle

$$A^{p,*}(M, \tilde{E}) = A^{0,*}(M, \mathbb{C}) \otimes A^{p,0}(M, \mathbb{C}) \otimes \tilde{E},$$

with splitting into forms with $*$ even and odd, respectively, see Example C.27.2. In this case, the canonical index form is

$$\operatorname{Td}(M) \wedge \operatorname{ch}(A^{p,0}(M, \mathbb{C}) \otimes \tilde{E}) = \operatorname{Td}(M) \wedge \operatorname{ch}(A^{p,0}(M, \mathbb{C})) \wedge \operatorname{ch}(\tilde{E}),$$

where Td and ch denote Todd genus and Chern character, see Theorem III.13.15 in [LM].

Before returning to the discussion of Kähler manifolds, we discuss a situation where a group G as above arises in a natural way. To that end, suppose in addition that M is simply connected. Let $F \to \tilde{M}$ be a Hermitian line bundle with Hermitian connection ∇ such that the curvature form ω of ∇ is Γ-invariant.

8.29 Lemma. *Let $p \in M$, $\gamma \in \Gamma$, and $u \colon F_p \to F_{\gamma p}$ be a unitary map. Then there is a unique lift of γ to an isomorphism g of F over γ with $g_p = u$ and such that g preserves Hermitian metric and connection of F.*

Sketch of proof. Let $q \in M$ and $y \in F_q$. Choose a piecewise smooth path c in M from p to q. Let x be the element in F_p which is parallel to y along c. Set $g(y) = z$, where z is parallel to ux along $\gamma \circ c$. The dependence of parallel translation on curvature together with the Γ-invariance of ω shows that z only depends on the homotopy class of c. Since M is simply connected, g is well defined. By definition, u preserves parallel translation, hence ∇. Since ∇ is Hermitian, g preserves the Hermitian metric of F. □

It follows from Lemma 8.29 that there is an extension

$$1 \to \mathrm{U}(1) \to G \to \Gamma \to 1 \tag{8.30}$$

of Γ by $U(1)$ which lifts the action of Γ on M to an action of G on F as in (8.24). Furthermore, the action of G preserves metric and connection of F.

We now return to the discussion of Kähler-hyperbolic Kähler manifolds. We let M be a closed and connected Kähler manifold and $E \to M$ be a flat vector bundle. We let $\tilde{M} \to M$ be the universal covering of M and $\tilde{E} \to M$ be the pull back of E.

8.31 Main Theorem. *If $\tilde{\omega} = d\eta$ with $\|\eta\|_\infty < \infty$, then*

$$\mathcal{H}_2^{p,q}(\tilde{M}, \tilde{E}) \neq 0 \quad \text{if } p + q = m.$$

Proof. Let $F = \tilde{M} \times \mathbb{C}$ be the trivial line bundle with canonical Hermitian metric and flat connection ∇^0, and consider the Hermitian connections

$$\nabla^t = \nabla^0 + it\eta.$$

The curvature form of ∇^t is $itd\eta = it\tilde{\omega}$, where $\tilde{\omega}$ denotes the Kähler form of $\tilde{M}$ with respect to the induced Kähler metric. Since $\tilde{\omega}$ is Γ-invariant, we get extensions G_t of Γ by $U(1)$ with induced actions on F as in Lemma 8.29 above. For each fixed t, consider the twist D^{∇^t} of the Dolbeault operator $D = \bar{\partial} + \bar{\partial}^*$ on $A^{p,*}(M, \tilde{E})$ by F with connection ∇^t. With respect to the splitting into forms with $*$ even and odd, the index form of $(D^{\nabla^t})^+$ is

$$\operatorname{Td}(M) \wedge \operatorname{ch}(A^{p,0}(M, \mathbb{C})) \wedge \operatorname{ch}(\tilde{E}) \wedge \operatorname{ch}(F),$$

where $\operatorname{ch} F = \exp(-t\omega/2\pi)$, see (A.44) and Remark 8.28 above. Hence the index form is a power series in t, where the coefficient of t^m in degree $\dim_{\mathbb{R}} M$ is

$$\frac{\operatorname{rk} E}{m!(-2\pi)^m}\binom{m}{p} \cdot \tilde{\omega}^m.$$

By Theorem 8.27, the index $\operatorname{ind}_{G_t}((D^{\nabla^t})^+)$ is the integral of the index form over M. It follows that $\operatorname{ind}_{G_t}((D^{\nabla^t})^+)$ is a polynomial in t, where the coefficient of t^m is non-zero[22]. Hence the zeros of this polynomial are isolated.

On the other hand, $\mathcal{H}_2^{p,q}(\tilde{M}, \tilde{E}) = 0$ would imply that $\operatorname{ind}_{G_t}(D^{\nabla^t}) = 0$ for all t sufficiently small, by Lemma 8.17. □

8.32 Theorem. *If $\tilde{\omega} = d\eta$ with $\|\eta\|_\infty < \infty$, then*

$$(-1)^m \chi(M) = \sum_{p+q=m} \dim_\Gamma \mathcal{H}_2^{p,q}(\tilde{M}, \tilde{E}) > 0,$$

where Γ denotes the fundamental group of M.

Proof. This is immediate from Theorems 8.21, 8.9, and 8.31, compare the proof of Theorem 8.22. □

[22] This is in contrast to the case of compact manifolds, where the index of elliptic differential operators is integral and would not depend on t.

9 Kodaira Embedding Theorem

Let M be a closed complex manifold of complex dimension m. Recall that a holomorphic line bundle $E \to M$ is positive or negative if its first Chern class $c_1(E)$ has a positive or negative representative, respectively, see Definition 5.59. It is immediate that a holomorphic line bundle is positive iff its dual is negative. In Examples 5.60 we showed that the tautological bundle $U \to \mathbb{C}P^m$ is negative and that the hyperplane bundle $H \to \mathbb{C}P^m$ is positive.

If M admits a positive holomorphic line bundle, then a positive representative of its first Chern class is a Kähler metric on M. In this sense the topic of this section belongs to Kähler geometry.

9.1 Remark. We say that a cohomology class in $H^2(M, \mathbb{C})$ is *integral* if it is in the image of the canonical morphism $H^2(M, \mathbb{Z}) \to H^2(M, \mathbb{C})$. If M is a Kähler manifold, then a cohomology class in $H^2(M, \mathbb{C})$ is the first Chern class of a holomorphic line bundle iff it is of type $(1, 1)$ and integral, see e.g. Proposition III.4.6 in [Wel]. It follows that a closed complex manifold M admits a positive holomorphic line bundle iff M has a Kähler metric g with integral Kähler class $[\omega]$. Then M is called a *Hodge manifold.*

Closed oriented surfaces are Hodge manifolds: Let S be such a surface and g be a Riemannian metric on S. Then the rotation by a positive right angle, in the tangent spaces of S, is a parallel complex structure on S and turns S into a Kähler manifold. Since S is of real dimension two, $H^2(S, \mathbb{R}) \cong \mathbb{R}$. Now the image of $H^2(S, \mathbb{Z})$ in $H^2(S, \mathbb{R})$ is a lattice, hence a proper rescaling of g has an integral Kähler class. Hence S is a Hodge manifold. By the latter argument we also get that a closed Kähler manifold M with first Betti number $b_1(M) = 1$ is a Hodge manifold.

The pull back of a positive holomorphic line bundle along a holomorphic immersion is positive. In particular, if M admits a holomorphic immersion into $\mathbb{C}P^n$ for some $n \geq m$, then M admits a positive holomorphic line bundle. Vice versa, Hodge conjectured and Kodaira proved that M admits a holomorphic embedding into $\mathbb{C}P^n$ for some $n \geq m$ if M has a positive line bundle. We present Kodaira's proof of his embedding theorem; our source is Section VI.4 in [Wel].

Kodaira's proof uses the following general construction which associates to a holomorphic line bundle $E \to M$ a holomorphic map from an open part M' of M to a complex projective space: Let $\mathcal{O}(M, E)$ be the space of holomorphic sections of E and M' be the open set of points $p \in M$ such that $\sigma(p) \neq 0$ for some $\sigma \in \mathcal{O}(M, E)$. Since M is compact, $\mathcal{O}(M, E)$ is of finite dimension. Consider the evaluation map

$$\varepsilon\colon E^* \to \mathcal{O}(M, E)^*, \quad \varepsilon(\varphi)(\sigma) = \varphi(\sigma(p)), \tag{9.2}$$

where p is the foot point of φ. Clearly ε is holomorphic. For $\varphi \in E^*$ non-zero, $\varepsilon(\varphi) \neq 0$ precisely for $p \in M'$. Hence we obtain a holomorphic map

$$k\colon M' \to P(\mathcal{O}(M, E)^*), \quad k(p) = [\varepsilon(\varphi)], \tag{9.3}$$

where φ is any non-zero element of the fiber of E^* over p.

There is a somewhat less abstract way of describing k. Let $B = (\sigma_0, \dots, \sigma_k)$ be a basis of $\mathcal{O}(M, E)$ and φ be a nowhere vanishing holomorphic section of E^* over an open subset U of M. Then the holomorphic map

$$k_B \colon U \to \mathbb{C}P^k, \quad k_B(p) = [\varphi \circ \sigma_0, \dots, \varphi \circ \sigma_k], \tag{9.4}$$

does not depend on the choice of φ. Hence k_B is well defined on M'. It is clear that k_B corresponds to k in the coordinates of $\mathcal{O}(M, E)^*$ defined by the basis B.

We say that a holomorphic line bundle $E \to M$ is *very ample* if, in the above construction, $M' = M$ and k (or k_B) is an embedding. We say that E is *ample* if some positive power of E is very ample.

9.5 Exercise. Show that E^* is isomorphic to the pull back under k of the tautological bundle over $P(\mathcal{O}(M, E)^*)$. Conclude that ample bundles are positive.

Kodaira proves the following more precise version of the embedding theorem.

9.6 Kodaira Embedding Theorem. *Let M be a closed complex manifold and $E \to M$ be a holomorphic line bundle. If E is positive, then E is ample.*

Let E be a holomorphic line bundle over M and p be a point in M. Let $\varphi \colon U \to E^*$ be a nowhere vanishing holomorphic section of E^*, where U is an open subset of M containing p. Then the 1-*jet* of a smooth section σ of E at p with respect to φ is

$$J^1_{\varphi,p}(\sigma) := ((\varphi \circ \sigma)(p), (\varphi \circ \sigma)'(p)) \in \mathbb{C} \oplus (T^*_p M \otimes \mathbb{C}), \tag{9.7}$$

where $(\varphi \circ \sigma)'$ denotes the derivative of $\varphi \circ \sigma$. The proof that positive holomorphic line bundles are ample relies on the following lemma.

9.8 Lemma. *Let E be a line bundle over M. Suppose that*

1) *for any two different points $p, q \in M$, the map*

$$J^0_{p,q} \colon \mathcal{O}(M, E) \to E_p \oplus E_q, \quad J^0_{p,q}(\sigma) = (\sigma(p), \sigma(q)),$$

is surjective and that

2) *for any point $p \in M$ and section φ about p as above, the map*

$$J^1_{\varphi,p} \colon \mathcal{O}(M, E) \to \mathbb{C} \oplus (T^*_p M \otimes \mathbb{C})$$

is surjective. Then E is very ample.

Proof. We first observe that $M' = M$, by Assumption 1). For a basis $B = (\sigma_0, \dots, \sigma_k)$ of $\mathcal{O}(M, E)$, we consider $k_B \colon M \to \mathbb{C}P^k$ as above.

Let p, q be different points of M. By Assumption 1), there are global holomorphic sections σ_p and σ_q of E with

$$\sigma_p(p) \neq 0, \; \sigma_p(q) = 0 \quad \text{and} \quad \sigma_q(q) \neq 0, \; \sigma_q(p) = 0.$$

Then σ_p and σ_q are linearly independent. If we choose B with $\sigma_0 = \sigma_p$ and $\sigma_1 = \sigma_q$, then clearly $k_B(p) \neq k_B(q)$. It follows that k_B is injective, for any choice of basis B.

Let p be a point in M and φ be a holomorphic section of E^* in an open neighborhood U of p as above. By Assumption 2), there are holomorphic sections $\sigma_0, \dots, \sigma_m$ in $\mathcal{O}(M, E)$ such that $J^1_{\varphi,p}(\sigma_0) = (1, 0)$ and $J^1_{\varphi,p}(\sigma_i) = (0, \beta_i)$, where $(\beta_1, \dots, \beta_m)$ is a basis of $T^*_p M \otimes \mathbb{C}$. Since $J^1_{\varphi,p}$ is a linear map, $\sigma_0, \dots, \sigma_m$ are linearly independent in $\mathcal{O}(M, E)$. Therefore we can complete them to a basis B of $\mathcal{O}(M, E)$. With respect to affine coordinates $(z_1, \dots, z_k)$ on $\{z_0 \neq 0\} \subset \mathbb{C}P^k$, we get

$$dk_B(p) = ((\beta_1, \dots, \beta_m, (\varphi \circ \sigma_{m+1})'(p), \dots, (\varphi \circ \sigma_k)'(p)).$$

Since $(\beta_1, \dots, \beta_m)$ are linearly independent, $dk_B(p)$ has rank m. Hence k_B is an immersion. Now M is compact, hence k_B is an embedding. □

9.1 Proof of the Embedding Theorem. We use notation and results from Subsection 3.4, in particular Examples 3.45 and Lemmas 3.46 and 3.47. We denote the blow up of M in a point p by M_p and use an index p for objects associated to M_p. For points $p \neq q$ in M, we let $M_{pq} = (M_p)_q$ and $\pi_{pq} \colon M_{pq} \to M$ be the canonical projection. There is a natural identification of M_{pq} with M_{qp} such that $\pi_{pq} = \pi_{qp}$. We let L_{pq} be the holomorphic line bundle associated to the hypersurface $\pi^{-1}_{pq}(p) \cup \pi^{-1}_{pq}(q)$.

9.9 Lemma. *Let E, F be holomorphic line bundles over M and suppose $E > 0$. Then given $k_0 \geq 1$ there is $n_0 \geq 0$ such that*

$$\pi_p^* E^n \otimes \pi_p^* F \otimes (L_p^*)^k > 0 \quad \text{and} \quad \pi_{pq}^* E^n \otimes \pi_{pq}^* F \otimes (L_{pq}^*)^k > 0$$

for all $p \neq q \in M$, $n \geq n_0$, and $k \in \{1, \dots, k_0\}$.

Proof. Choose a Hermitian metric on E such that the curvature of its Chern connection satisfies $i\Theta_E > 0$, see Lemma 5.61. Fix a Hermitian metric on F and denote the curvature of its Chern connection by Θ_F. The Chern connection of the tensor product $E_1 \otimes E_2$ of two Hermitian holomorphic vector bundles $E_1, E_2 \to M$ with induced Hermitian metric

$$(e_1 \otimes e_2, e_1' \otimes e_2') = (e_1, e_1')\,(e_2, e_2')$$

is given by the usual product rule, hence its curvature is the sum of the curvatures of the Chern connections on E_1 and E_2, $\Theta = \Theta_1 + \Theta_2$. Since M is compact and $i\Theta_E > 0$, there is an $n_0 > 0$ such that $i(n_0\Theta_E + \Theta_F) > 0$. In particular, $E^n \otimes F > 0$ for all $n \geq n_0$.

Let $p \in M$ and consider the projection $\pi_p \colon M_p \to M$. The Chern connection of the pull back of a Hermitian metric on the pull back of a holomorphic vector bundle is the pull back of the Chern connection of the original Hermitian metric on the original bundle. Hence the Chern connection of the pull back metric on $\pi_p^* E^n \otimes \pi_p^* F$ has curvature Θ_n equal to the pull back of the curvature on $E^n \otimes F$. Therefore $i\Theta_n(v, Jv) > 0$ for v not in the kernel of π_{p*}, that is, except for $v = 0$ or v tangent to S_p.

Positive powers of L_p^* will bring positivity everywhere. To get n_0 independent of the chosen point $p \in M$, we proceed as follows: Choose holomorphic coordinates $z\colon U \to U'$ on M, where U' contains the ball of radius 4 about 0 in $\mathbb{C}^m$. Set

$$U_i = \{p \in U \mid |z(p)| < i\}.$$

Fix a cut off function χ on M with $\chi = 1$ on U_2 and $\chi = 0$ on $M \setminus U_3$.

For p in the closure $\overline{U}_1$ of U_1, we use $z - z(p)$ as holomorphic coordinates in the construction of the blow up M_p. Via the canonical projection π_p, we identify $M_p \setminus \pi_p^{-1}(\overline{U}_1)$ with $M \setminus \overline{U}_1$. In this sense, we view L_p^* restricted to $M_p \setminus \pi_p^{-1}(\overline{U}_1)$ as a holomorphic line bundle over $M \setminus \overline{U}_1$. The holomorphic section Φ_0^* dual to the holomorphic section Φ_0 as in Example 3.45.4 turns it into a trivial bundle. We define a Hermitian metric h_p' on this part of L_p^* by associating length 1 to Φ_0^*. Then Φ_0^* is parallel with respect to the Chern connection of h_p', and hence the curvature of the latter vanishes identically.

Since the hyperplane bundle $H \to \mathbb{C}P^{m-1}$ is positive, there is a Hermitian metric h on H such that the curvature of its Chern connection satisfies $i\Theta_h > 0$. Over $\pi_p^{-1}(U_4)$, we have $L_p^* = \sigma_p^* H$, where $\sigma_p\colon \pi_p^{-1}(U_4) \to \mathbb{C}P^{m-1}$ is the canonical projection. Therefore we view $h_p'' = \sigma_p^* h$ as a Hermitian metric on that part of L_p^*. The curvature Θ_p'' of the Chern connection of h_p'' is the pull back of Θ_h, and hence $i\Theta_p'' \geq 0$ and $i\Theta_p'' > 0$ on the tangent bundle of S_p. We now let

$$h_p = (\chi \circ \pi_p) h_p'' + (1 - \chi \circ \pi_p) h_p'.$$

Then the curvature Θ_p of the Chern connection of h_p vanishes on $M_p \setminus \pi_p^{-1}(U_3)$ and $\Theta_p = \Theta_p''$ on $\pi^{-1}(U_2)$. Since $\sigma_p(q)$ is smooth in $p \in \overline{U}_1$ and $q \in \overline{U}_3 \setminus U_2$, the family of Hermitian metrics h_p'', $p \in \overline{U}_1$, is uniformly bounded on $\overline{U}_3 \setminus U_2$, together with derivatives of any order. It follows that Θ_p is uniformly bounded on $\overline{U}_3 \setminus U_2$, independently of $p \in \overline{U}_1$.

The curvature of the Chern connection of the induced Hermitian metric on $E^n \otimes F \otimes (L^*)^k$ is $\Theta = \Theta_n + k\Theta_p$. Clearly $i\Theta$ is positive outside $\overline{U}_3 \setminus U_2$. If we increase the above n_0 if necessary, the positivity of $i\Theta_n$ outweighs the possible negativity of $ki\Theta_p$ on $\overline{U}_3 \setminus U_2$. By the uniform boundedness of Θ_p, n_0 can be chosen independently of $p \in \overline{U}_1$. Now M is compact, hence can be covered by a finite number of sets U_1 as in the argument.

The argument in the case of two points $p \neq q$ in a set U_1 as above is similar. We use the same construction, but now with two pull backs of H corresponding to the two points p and q. For p and q uniformly far apart, we can apply the argument for the case of one point twice. □

We now come to the final steps of the proof of the embedding theorem. We will assume some elementary facts from sheaf theory. For a sheaf $\mathcal{F}$ over M and an open subset U of M, we denote by $\mathcal{F}(U)$ the space of sections of $\mathcal{F}$ over U. For $p \in M$, we denote by $\mathcal{F}_p$ the stalk of $\mathcal{F}$ at p.

Let $\mathcal{O}$ be the *structure sheaf* of M, that is, the sheaf of rings of germs of holomorphic functions on M. For $p \in M$, let $\mathcal{I}_p$ and $\mathcal{I}_p^2$ be the subsheaves of ideals in $\mathcal{O}$ of germs of holomorphic functions which vanish at p respectively vanish at p at least of second order. Then we have a short exact sequence

$$0 \to \mathcal{I}_p^2 \to \mathcal{O} \to \mathcal{O}/\mathcal{I}_p^2 \to 0. \tag{9.10}$$

For $q \neq p$, the stalk $(\mathcal{O}/\mathcal{I}_p^2)_q = 0$. Moreover, the map

$$J_p^1 \colon \mathcal{O}_p \to \mathbb{C} \oplus (T_p^*M \otimes \mathbb{C}), \quad J_p^1(f) = (f(p), f'(p)), \tag{9.11}$$

induces an isomorphism

$$(\mathcal{O}/\mathcal{I}_p^2)_p \to \mathbb{C} \oplus (T_p^*M \otimes \mathbb{C}). \tag{9.12}$$

Let $F \to M$ be a holomorphic line bundle. Denote by $\mathcal{O}(F)$ the sheaf of germs of holomorphic sections of F. Since $\mathcal{O}(F)$ is a locally free sheaf of modules over the structure sheaf $\mathcal{O}$ of M, we obtain an induced short exact sequence

$$0 \to \mathcal{O}(F) \otimes \mathcal{I}_p^2 \to \mathcal{O}(F) \to \mathcal{O}(F) \otimes (\mathcal{O}/\mathcal{I}_p^2) \to 0, \tag{9.13}$$

where the tensor product is taken over $\mathcal{O}$. We may view $\mathcal{O}(F) \otimes \mathcal{I}_p^2$ as sheaf of germs of holomorphic sections of F which vanish at least of second order at the distinguished point p. The stalks in $q \neq p$ are

$$(\mathcal{O}(F) \otimes \mathcal{I}_p^2)_q = \mathcal{O}(F)_q \quad \text{and} \quad (\mathcal{O}(F) \otimes (\mathcal{O}/\mathcal{I}_p^2))_q = \{0\}. \tag{9.14}$$

To identify the stalk of $\mathcal{O}(F) \otimes (\mathcal{O}/\mathcal{I}_p^2)$ at p, we observe that the map $J_{\varphi,p}^1$ from Lemma 9.8 is also defined on the stalk $\mathcal{O}(F)_p$ and that it induces an isomorphism

$$(\mathcal{O}(F) \otimes (\mathcal{O}/\mathcal{I}_p^2))_p \to \mathbb{C} \oplus (T_p^*M \otimes \mathbb{C}). \tag{9.15}$$

9.16 Lemma. *Let $E \to M$ be a positive holomorphic line bundle. In Lemma* 9.9, *let F be the canonical bundle of M, $F = K$, and $k_0 = m + 1$. Then E^n, $n \geq n_0$, satisfies the assumptions of Lemma* 9.8 *and hence E is ample.*

Proof. The proof of Assumption 1) in Lemma 9.8 is similar to and easier than the proof of Assumption 2). Therefore we concentrate on Assumption 2) and leave the proof of Assumption 1) to the reader.

Let $p \in M$. By (9.14), the restriction map induces an isomorphism from the space $H^0(M, \mathcal{O}(E^n) \otimes (\mathcal{O}/\mathcal{I}_p^2))$ of global holomorphic sections of $\mathcal{O}(E^n) \otimes (\mathcal{O}/\mathcal{I}_p^2)$ onto the stalk $(\mathcal{O}(E^n) \otimes (\mathcal{O}/\mathcal{I}_p^2))_p$. Hence it remains to show that

$$\mathcal{O}(M, E^n) = H^0(M, \mathcal{O}(E^n)) \to H^0(M, \mathcal{O}(E^n) \otimes (\mathcal{O}/\mathcal{I}_p^2)) \tag{9.17}$$

is surjective, by (9.15). Since points in M have codimension m and do not correspond to line bundles if $m > 1$, we will need to pass to blow ups of M to show the surjectivity in (9.17).

Let $\tilde{M} = M_p$ be the blow up of M in p. Let $\pi : \tilde{M} \to M$ be the canonical projection and $\tilde{E} = \pi^* E$. Denote by $\tilde{\mathcal{O}}$ the structure sheaf of $\tilde{M}$ and by $\mathcal{I}_S$ and $\mathcal{I}_S^2$ the subsheaves of $\tilde{\mathcal{O}}$ of germs of holomorphic functions which vanish along S respectively vanish at least of second order along S. Composition with π induces injective morphisms $\pi^* \colon \mathcal{O} \to \tilde{\mathcal{O}}$ and $\pi_\mu^* \colon \mathcal{O}(E^n) \to \mathcal{O}(\tilde{E}^n)$. Now E^n is trivial in a neighborhood of p in M, and the pull back of a local trivialization defines a trivialization of $\tilde{E}^n$ in a neighborhood of S. From the coordinate description of $\tilde{M}$ in Subsection 2.4 it is then clear that a holomorphic section σ of E^n vanishes of at least second order at p iff $\sigma \circ \pi$ vanishes of at least second order about S. It follows that π induces morphisms

$$\pi_\lambda^* \colon \mathcal{O}(E^n) \otimes \mathcal{I}_p^2 \to \mathcal{O}(\tilde{E}^n) \otimes \mathcal{I}_S^2$$

and

$$\pi_\rho^* \colon \mathcal{O}(E^n) \otimes (\mathcal{O}/\mathcal{I}_p^2) \to \mathcal{O}(\tilde{E}^n) \otimes (\tilde{\mathcal{O}}/\mathcal{I}_S^2).$$

Hence we get a commutative diagram

$$\begin{array}{ccccccccc} 0 & \longrightarrow & \mathcal{O}(\tilde{E}^n) \otimes \mathcal{I}_S^2 & \longrightarrow & \mathcal{O}(\tilde{E}^n) & \longrightarrow & \mathcal{O}(\tilde{E}^n) \otimes \tilde{\mathcal{O}}/\mathcal{I}_S^2 & \longrightarrow & 0 \\ & & \uparrow{\scriptstyle \pi_\lambda^*} & & \uparrow{\scriptstyle \pi_\mu^*} & & \uparrow{\scriptstyle \pi_\rho^*} & & \\ 0 & \longrightarrow & \mathcal{O}(E^n) \otimes \mathcal{I}_p^2 & \longrightarrow & \mathcal{O}(E^n) & \longrightarrow & \mathcal{O}(E^n) \otimes \mathcal{O}/\mathcal{I}_p^2 & \longrightarrow & 0 \end{array} \tag{9.18}$$

of short exact sequences of sheaves.

Clearly π_λ^* and π_μ^* are injective. We claim that they are also surjective. This is clear in the case $m = 1$ since then blowing up is trivial and π is biholomorphic. For $m > 1$, a section $\tilde{\sigma}$ of $\tilde{E}^n$ over $\tilde{M}$ restricts to a section σ_p of $\tilde{E}^n$ over $\tilde{M} \setminus S = M \setminus \{p\}$. By Hartogs' theorem [GH, page 7], σ_p extends to a holomorphic section σ of E^n over M. By the continuity of σ and $\tilde{\sigma}$, we have $\pi_\mu^* \sigma = \tilde{\sigma}$. Moreover, if $\tilde{\sigma}$ vanishes of order at least 2 along S, then also σ at p. Hence π_λ^* and π_μ^* are surjective, hence isomorphic. The five lemma implies that π_ρ^* is an isomorphism as well.

A fortiori, the vertical morphisms in (9.18) induce isomorphisms between the corresponding cohomology groups in the long exact sequences associated to the horizontal short exact sequences. Thus by (9.17), it remains to show that

$$H^0(\tilde{M}, \mathcal{O}(\tilde{E}^n)) \to H^0(\tilde{M}, \mathcal{O}(\tilde{E}^n) \otimes (\tilde{\mathcal{O}}/\mathcal{I}_S^2))$$

is surjective. By the long exact sequence, this follows if $H^1(\tilde{M}, \mathcal{O}(\tilde{E}^n) \otimes \mathcal{I}_S^2) = 0$.

To show this we observe first that $\mathcal{I}_S = \mathcal{O}(L^*)$. In the notation of Example 3.45.4, this follows in W_i, $i \geq 1$, since there $f \in \mathcal{I}_S$ can be expressed as $f = f_i z_i$ with f_i holomorphic. That is, the functions z_i serve as a frame over W_i and $f_i z_i = f = f_j z_j$ over $W_i \cap W_j$. Now $\tilde{E}^n \otimes \tilde{K}^* \otimes (L^*)^2 > 0$ by Lemmas 3.46 and 3.47, hence

$$H^1(\tilde{M}, \mathcal{O}(\tilde{E}^n \otimes (L^*)^2)) = 0$$

by the Kodaira vanishing theorem 5.65. □

9.2 Two Applications. The Kodaira embedding theorem has many applications. We discuss two immediate ones.

9.19 Corollary. *Let M be a closed complex manifold and M_p be the blow up of M in a point $p \in M$. If M admits a holomorphic embedding into a complex projective space, then M_p as well.*

Proof. Suppose $f\colon M \to \mathbb{C}P^n$ is a holomorphic embedding. Then the pull back $E \to M$ of the hyperplane bundle H over $\mathbb{C}P^n$ is positive. By Lemma 9.9, $\pi_p^* E \otimes (L^*)^k$ is positive for k sufficiently large, where $\pi_p\colon M_p \to M$ is the natural map. Hence M_p admits a holomorphic embedding into some complex projective space, by Theorem 9.6. □

9.20 Corollary. *Let M be a closed complex manifold and $\tilde{M} \to M$ be a finite covering of M. If M admits a holomorphic embedding into a complex projective space, then $\tilde{M}$ as well.*

Proof. Suppose $f\colon M \to \mathbb{C}P^n$ is a holomorphic embedding. Then the pull back $E \to M$ of the hyperplane bundle H over $\mathbb{C}P^n$ is positive, therefore the pull back $\tilde{E} \to \tilde{M}$ of E is positive as well. Hence $\tilde{M}$ admits a holomorphic embedding into some complex projective space, by Theorem 9.6. □

9.21 Remark. For a finite covering $\tilde{M} \to M$ as above, averaging leads to the converse assertion: If $\tilde{M}$ admits a holomorphic embedding into a complex projective space, then also M, see [GH, page 192].

To complete the picture, we quote the theorem of Chow which says that a complex analytic submanifold of a complex projective space is a smooth projective variety, see for example [GH, page 167].

Appendix A Chern–Weil Theory

In this appendix, no Kähler classes are in the way and, therefore, connection and curvature forms are denoted by the more standard ω and Ω, respectively.

Let $V_1, \dots, V_k$ and V be vector spaces over the field $\mathbb{F} \in \{\mathbb{R}, \mathbb{C}\}$ and

$$\Phi\colon V_1 \times \dots \times V_k \to V \tag{A.1}$$

be a k-linear map. Let $\Lambda^*_{\mathbb{F}}\mathbb{R}^n$ be the algebra of alternating forms on $\mathbb{R}^n$ with values in $\mathbb{F}$. The map

$$(\varphi_1, v_1, \dots, \varphi_k, v_k) \mapsto (\varphi_1 \wedge \dots \wedge \varphi_k) \otimes \Phi(v_1, \dots, v_k), \tag{A.2}$$

where $\varphi_i \in \Lambda^*_{\mathbb{F}}\mathbb{R}^n$ and $v_i \in V_i$, is linear in each of its $2k$ arguments. Therefore it gives rise to a k-linear map

$$\Phi_\Lambda\colon (\Lambda^*_{\mathbb{F}}\mathbb{R}^n \otimes V_1) \times \dots \times (\Lambda^*_{\mathbb{F}}\mathbb{R}^n \otimes V_k) \to \Lambda^*_{\mathbb{F}}\mathbb{R}^n \otimes V. \tag{A.3}$$

We view $\Lambda^*_{\mathbb{F}}\mathbb{R}^n \otimes V_i$ and $\Lambda^*_{\mathbb{F}}\mathbb{R}^n \otimes V$ as the space of alternating forms on $\mathbb{R}^n$ with values in V_i and V, respectively. If $\alpha_1, \dots, \alpha_k$ are alternating forms on $\mathbb{R}^n$ with values in $V_1, \dots, V_k$ and degrees $d_1, \dots, d_k$, respectively, and if we express $\alpha_i = \sum \varphi_{ij} \otimes v_{ij}$ with $\varphi_{ij} \in \Lambda^*_{\mathbb{F}}\mathbb{R}^n$ and $v_{ij} \in V_i$, then

$$\Phi_\Lambda(\alpha_1, \dots, \alpha_k) = \sum_{(j_1,\dots,j_k)} (\varphi_{1j_1} \wedge \dots \wedge \varphi_{kj_k}) \otimes \Phi(v_{1j_1}, \dots, v_{kj_k}). \tag{A.4}$$

The degree of $\Phi_\Lambda(\alpha_1, \dots, \alpha_k)$ is $d = d_1 + \dots + d_k$. For $x_1, \dots, x_d \in \mathbb{R}^n$, we have

$$\begin{aligned}&\Phi_\Lambda(\alpha_1, \dots, \alpha_k)(x_1, \dots, x_d)\\ &= \frac{1}{d_1! \dots d_k!} \sum_\sigma \varepsilon(\sigma) \Phi(\alpha_1(x_{\sigma(1)}, \dots, x_{\sigma(d_1)}), \dots, \alpha_k(x_{\sigma(d-d_k+1)}, \dots, x_{\sigma(d)})).\end{aligned} \tag{A.5}$$

This is immediate from (A.4) and the definition of the wedge product. It is also clear from this formula that we are discussing a generalization of the wedge product introduced in (1.15).

Let G be a Lie group, and suppose that G acts linearly on the vector spaces $V_1, \dots, V_k$, and V (via given representations, which we suppress in the notation). Then G acts also on $\Lambda^*_{\mathbb{F}}\mathbb{R}^n \otimes V$,

$$g(\varphi \otimes v) := \varphi \otimes (gv), \tag{A.6}$$

and similarly for $\Lambda^*_{\mathbb{F}}\mathbb{R}^n \otimes V_i$. We say that Φ (from (A.1)) is *equivariant* (with respect to these G-actions) if

$$\Phi(gv_1, \dots, gv_k) = g\Phi(v_1, \dots, v_k) \tag{A.7}$$

for all $g \in G$ and $v_i \in V_i$, $1 \le i \le k$. We say that Φ is *invariant* if it is equivariant with respect to the trivial action on V. That is, if (A.7) holds without the factor g on the right hand side. If Φ is equivariant or invariant, then the induced map Φ_Λ is equivariant or invariant as well. This is immediate from (A.4) and (A.6).

A.8 Examples. 1) Via the adjoint representation, G acts on its Lie algebra $\mathfrak{g}$. The Lie bracket $[A, B]$ defines an equivariant bilinear map $\lambda\colon \mathfrak{g} \times \mathfrak{g} \to \mathfrak{g}$.

2) The adjoint representation of the general linear group $G = \mathrm{Gl}(m, \mathbb{F})$ on its Lie algebra $\mathfrak{g} = \mathfrak{gl}(m, \mathbb{F}) = \mathbb{F}^{m\times m}$ is given by conjugation, $\mathrm{Ad}_g A = gAg^{-1}$. The product AB of matrices defines an equivariant bilinear map $\mu\colon \mathfrak{g} \times \mathfrak{g} \to \mathfrak{g}$. The trace $\mathrm{tr}\colon \mathfrak{g} \to \mathbb{F}$ is an invariant linear map since $\mathrm{tr}(gAg^{-1}) = \mathrm{tr}\, A$ for all $g \in G$ and $A \in \mathfrak{g}$. Via polarization, the determinant gives rise to an invariant m-linear map $\mathfrak{g} \times \cdots \times \mathfrak{g} \to \mathbb{F}$, see below.

Let $P \to M$ be a principal bundle with structure group G. Let $E = P \times_G V$ and $E_i = P \times_G V_i$ be the vector bundles associated to the representations of G on V and V_i, $1 \le i \le k$. Recall that E consists of equivalence classes of tuples $[f, v]$ with relation $[fg, v] = [f, gv]$ for all $f \in P$, $g \in G$, and $v \in V$, and similarly for E_i. Suppose that Φ is equivariant. Then Φ induces a field of morphisms

$$\begin{aligned} E_1 \times \cdots \times E_k &\to E, \\ ([f, v_1], \dots, [f, v_k]) &\mapsto [f, \Phi(v_1, \dots, v_k)]. \end{aligned} \tag{A.9}$$

Let $\alpha_i \in \mathcal{A}^*(M, E_i)$ be differential forms on M with values in E_i, $1 \le i \le k$. With respect to a local section $\tau\colon U \to P$ of P, we can write (or identify)

$$\alpha_i = \sum \varphi_{ij} \otimes [\tau, v_{ij}] = \sum [\tau, \varphi_{ij} \otimes v_{ij}], \tag{A.10}$$

where the φ_{ij} are differential forms on U with values in $\mathbb{F}$ and the v_{ij} are smooth maps from U to V_i. We call $(\alpha_i)_\tau = \sum \varphi_{ij} \otimes v_{ij}$ the *principal part* of α_i with respect to τ. If Φ is equivariant, then $\Phi_\wedge$ is equivariant as well, and we obtain an induced field of morphisms on differential forms,

$$\begin{aligned} \mathcal{A}^*(M, E_1) \times \cdots \times \mathcal{A}^*(M, E_k) &\to \mathcal{A}^*(M, E), \\ (\alpha_1, \dots, \alpha_k) &\mapsto [\tau, \Phi_\wedge((\alpha_1)_\tau, \dots, (\alpha_k)_\tau)] \\ &= \sum [\tau, (\varphi_{1j_1} \wedge \cdots \wedge \varphi_{kj_k}) \otimes \Phi(v_{1j_1}, \dots, v_{kj_k})], \end{aligned} \tag{A.11}$$

where the degree is additive in the degrees of the arguments. In the case where $V = \mathbb{F}$ with trivial action of G, we view $\mathcal{A}^*(M, E) = \mathcal{A}^*(M, \mathbb{F})$, the space of differential forms on M with values in $\mathbb{F}$.

Although some of the following holds in greater generality, we now specialize to the case where $V_1 = \cdots = V_k = \mathfrak{g}$ with the adjoint action of G and $V = \mathbb{F}$ with the trivial G-action. That is, we consider an invariant k-linear map

$$\Phi\colon \mathfrak{g} \times \cdots \times \mathfrak{g} \to \mathbb{F}. \tag{A.12}$$

Recall that for $A, B \in \mathfrak{g}$,

$$\lambda(A, B) := (\mathrm{Ad}(e^{tA})B)'_{t=0} = [A, B]. \tag{A.13}$$

The induced map on alternating forms associates to alternating forms α and β with values in $\mathfrak{g}$ an alternating form

$$\lambda_\Lambda(\alpha, \beta) = \alpha \wedge_\lambda \beta \tag{A.14}$$

with values in $\mathbb{F}$, where the notation is from (A.3) on the left hand side and from (1.15) on the right, compare also Exercise 1.16. On decomposable forms $\alpha = \varphi \otimes A$ and $\beta = \psi \otimes B$ with $A, B \in \mathfrak{g}$,

$$\alpha \wedge_\lambda \beta = (\varphi \wedge \psi) \otimes [A, B]. \tag{A.15}$$

A.16 Lemma. *For $A, B_1, \dots, B_k \in \mathfrak{g}$,*

$$\sum_i \Phi(B_1, \dots, A \wedge_\lambda B_i, \dots, B_k) = 0.$$

Proof. Since Φ is invariant,

$$\Phi(\operatorname{Ad}(e^{tA})B_1, \dots, \operatorname{Ad}(e^{tA})B_k) = \Phi(B_1, \dots, B_k).$$

Differentiation of this equation yields the claim. □

A.17 Corollary. *Let $\alpha, \beta_1, \dots, \beta_k$ be alternating forms with values in $\mathfrak{g}$ and of degrees $d, d_1, \dots, d_k$, respectively. Then, with $s(i) := (d_1 + \dots + d_{i-1})d$,*

$$\sum_i (-1)^{s(i)} \Phi_\Lambda(\beta_1, \dots, \alpha \wedge_\lambda \beta_i, \dots, \beta_k) = 0.$$

Proof. By multilinearity, it is sufficient to consider decomposable forms. For these, the claim is immediate from (A.4) and Lemma A.16. The sign arises from moving the differential form part of α to the leading position. □

We return to the principal bundle $P \to M$ with structure group G and let D be a connection on P. We interpret the curvature R of D as a differential two-form on M with values in $P \times_G \mathfrak{g}$. With respect to a local section $\tau\colon U \to P$, we denote the principal parts of D and R by ω and Ω, differential one- and two-forms on U with values in $\mathfrak{g}$. By the above, we obtain a differential $2k$-form $\Phi(P, R)$ on M with values in $\mathbb{F}$ by setting

$$\Phi(P, R) := [\tau, \Phi_\Lambda(\Omega, \dots, \Omega)]. \tag{A.18}$$

A.19 Fundamental Lemma. *For the curvature R of a connection D on P, the differential form $\Phi(P, R)$ is closed and its cohomology class in $H^{2k}(M, \mathbb{F})$ does not depend on the choice of D.*

Proof. Fix a local section τ of P. In the notation of (A.14) and (1.15), the Bianchi identity says $d\Omega = \Omega \wedge_\lambda \omega$, where λ is the Lie bracket of $\mathfrak{g}$. By Exercise 1.16, $\Omega \wedge_\lambda \omega = -\omega \wedge_\lambda \Omega$. Hence

$$\begin{aligned} d(\Phi(P, R)) &= \sum [\tau, \Phi_\Lambda(\Omega, \dots, d\Omega, \dots, \Omega)] \\ &= \sum [\tau, \Phi_\Lambda(\Omega, \dots, \Omega \wedge_\lambda \omega, \dots, \Omega)] = 0, \end{aligned}$$

by (A.17). This shows the first assertion.

Let D_0 and D_1 be connections on P. Then $D_t = (1-t)D_0 + tD_1$ is also a connection on P. The principal part of D_t with respect to τ is

$$\omega_t = (1-t)\omega_0 + t\omega_1 = \omega_0 + t\beta,$$

where the difference $\beta = \omega_1 - \omega_0$ is a one-form with values in $\mathfrak{g}$. We recall the structure equation

$$\Omega_t = d\omega_t + \frac{1}{2}\omega_t \wedge_\lambda \omega_t.$$

By Exercise 1.16, $\omega_t' \wedge_\lambda \omega_t = \omega_t \wedge_\lambda \omega_t'$, where the prime indicates differentiation with respect to t. Hence

$$\begin{aligned}\Omega_t' &= d\omega_t' + \frac{1}{2}\omega_t' \wedge_\lambda \omega_t + \frac{1}{2}\omega_t \wedge_\lambda \omega_t' \\ &= d\omega_t' + \omega_t' \wedge_\lambda \omega_t = d\beta + \beta \wedge_\lambda \omega_t.\end{aligned}$$

With the Bianchi identity and Corollary A.17,

$$\begin{aligned}
&d\Big(\sum_i \Phi_\Lambda(\Omega_t, \dots, \beta, \dots, \Omega_t)\Big) \\
&\quad = \sum_{j<i} \Phi_\Lambda(\Omega_t, \dots, d\Omega_t, \dots, \beta, \dots, \Omega_t) + \sum_i \Phi_\Lambda(\Omega_t, \dots, d\beta, \dots, \Omega_t) \\
&\qquad - \sum_{j>i} \Phi_\Lambda(\Omega_t, \dots, \beta, \dots, d\Omega_t, \dots, \Omega_t) \\
&\quad = \sum_{j<i} \Phi_\Lambda(\Omega_t, \dots, \Omega_t \wedge_\lambda \omega_t, \dots, \beta, \dots, \Omega_t) \\
&\qquad - \sum_i \Phi_\Lambda(\Omega_t, \dots, \beta \wedge_\lambda \omega_t, \dots, \Omega_t) \\
&\qquad - \sum_{j>i} \Phi_\Lambda(\Omega_t, \dots, \beta, \dots, \Omega_t \wedge_\lambda \omega_t, \dots, \Omega_t) \\
&\qquad + \sum_i \Phi_\Lambda(\Omega_t, \dots, \Omega_t', \dots, \Omega_t) \\
&\quad = (\Phi_\Lambda(\Omega_t, \dots, \Omega_t))',
\end{aligned}$$

the principal part of $\Phi(P, R_t)'$. Hence

$$\Phi(P, R_1) - \Phi(P, R_0) = dT, \tag{A.20}$$

where T is the *transgression form*,

$$T = [\tau, \sum_i \int_0^1 \Phi_\Lambda(\Omega_t, \dots, \beta, \dots, \Omega_t)dt]. \tag{A.21}$$

Note that β is the principal part of $B = D_1 - D_0$, a one-form on M with values in $P \times_G \mathfrak{g}$. Hence T is well defined on M. □

Since the curvature form of a connection D has degree two, $\Phi(P, R) = 0$ if there is a pair $i < j$ such that Φ is alternating in the corresponding variables,

$$\Phi(A_1, \dots, A_i, \dots, A_j, \dots, A_m) = -\Phi(A_1, \dots, A_j, \dots, A_i, \dots, A_m).$$

In the applications of Theorem A.19, we are therefore only interested in the case where Φ is symmetric and invariant.

Let $S_k(\mathfrak{g})$ be the space of symmetric k-linear forms on $\mathfrak{g}$ with values in $\mathbb{F}$. Let $\mathbb{F}[\mathfrak{g}]$ be the algebra of polynomials on $\mathfrak{g}$ with values in $\mathbb{F}$ and $\mathbb{F}_k[\mathfrak{g}]$ be the subspace of homogeneous polynomials of degree k. The evaluation along the diagonal gives rise to an isomorphism

$$S_k(\mathfrak{g}) \to \mathbb{F}_k[\mathfrak{g}], \; \tilde{\Phi} \mapsto \Phi, \quad \text{where } \Phi(A) := \tilde{\Phi}(A, \dots, A). \tag{A.22}$$

The inverse of this isomorphism goes under the name polarization. Since the isomorphism is equivariant with respect to the natural G-actions, G-invariant symmetric multilinear forms correspond exactly to G-invariant polynomials.

In terms of a given basis $(B_1, \dots, B_m)$ of $\mathfrak{g}$, a homogeneous polynomial Φ of degree k on $\mathfrak{g}$ can be written as

$$\Phi(A) = \sum_{\mu} f_\mu a_{\mu(1)} \dots a_{\mu(k)}, \tag{A.23}$$

where μ runs over all non-decreasing maps $\mu\colon \{1, \dots, k\} \to \{1, \dots, m\}$, the coefficients f_μ are in $\mathbb{F}$ and the a_j are the coordinates of A with respect to the chosen basis, $A = \sum a_j B_j$. Then the polarization of Φ is

$$\tilde{\Phi}(A_1, \dots, A_k) = \frac{1}{k!} \sum_{\mu,\sigma} f_\mu a_{\sigma(1)\mu(1)} \dots a_{\sigma(k)\mu(k)}, \tag{A.24}$$

where now, in addition, σ runs over all permutations of $\{1, \dots, k\}$ and the a_{ij} are the coordinates of the vector A_i. By multilinearity and (A.23),

$$\tilde{\Phi}(A, \dots, A) = \sum_{j_1,\dots,j_k} \tilde{\Phi}(B_{j_1}, \dots, B_{j_k}) a_{j_1} \dots a_{j_k} = \sum_{\mu} f_\mu a_{\mu(1)} \dots a_{\mu(k)}. \tag{A.25}$$

We let $\Lambda_{\mathbb{F}}^{\text{even}} \mathbb{R}^n := \oplus \Lambda_{\mathbb{F}}^{2i} \mathbb{R}^n$ and consider

$$\alpha = \sum \varphi_j \otimes B_j \in \Lambda_{\mathbb{F}}^{\text{even}} \mathbb{R}^n \otimes \mathfrak{g}. \tag{A.26}$$

Since $\Lambda_{\mathbb{F}}^{\text{even}} \mathbb{R}^n$ is a commutative algebra, (A.25) implies

$$\begin{aligned} \tilde{\Phi}_\Lambda(\alpha, \dots, \alpha) &= \sum_{j_1,\dots,j_k} \tilde{\Phi}(B_{j_1}, \dots, B_{j_k}) \varphi_{j_1} \wedge \dots \wedge \varphi_{j_k} \\ &= \sum f_\mu \varphi_{\mu(1)} \wedge \dots \wedge \varphi_{\mu(k)} =: \Phi_\Lambda(\alpha) \in \Lambda_{\mathbb{F}}^{\text{even}} \mathbb{R}^n. \end{aligned} \tag{A.27}$$

The definition of $\Phi_\Lambda(\alpha)$ is very natural, and we could have started with it. On the other hand, the detour over polarization shows that we are in the framework of what we discussed before.

Equation A.27 implies that for homogeneous polynomials Φ and Ψ,

$$(\Phi \cdot \Psi)_\Lambda(\alpha) = \Phi_\Lambda(\alpha) \wedge \Psi_\Lambda(\alpha). \tag{A.28}$$

In particular, linear extension to the algebra $\mathbb{F}[\mathfrak{g}]$ of polynomials gives rise to an algebra morphism

$$\mathbb{F}[\mathfrak{g}] \to \Lambda^{\mathrm{even}}_{\mathbb{F}}\mathbb{R}^n, \ \ \Phi \mapsto \Phi_\Lambda(\alpha). \tag{A.29}$$

If α is a sum of forms of strictly positive (and even) degrees, then the natural extension of this map from $\mathbb{F}[\mathfrak{g}]$ to the larger algebra $\mathbb{F}[[\mathfrak{g}]]$ of formal power series in $\mathfrak{g}$ over $\mathbb{F}$ is well defined. For this we note that $\Phi_\Lambda(\alpha)$ has degree kd if Φ is homogeneous of degree k and α is of pure degree d and that alternating forms on $\mathbb{R}^n$ of degree $> n$ vanish. In both cases, $\mathbb{F}[\mathfrak{g}]$ or $\mathbb{F}[[\mathfrak{g}]]$, if Φ is invariant under the natural action of G, then also $\Phi_\Lambda(\alpha)$.

We return again to the principal bundle P with structure group G. Denote by $\mathbb{F}[\mathfrak{g}]^G \subset \mathbb{F}[\mathfrak{g}]$ and $\mathbb{F}[[\mathfrak{g}]]^G \subset \mathbb{F}[[\mathfrak{g}]]$ the subalgebra of G-invariant polynomials and G-invariant formal power series, respectively. Let $\alpha \in \mathcal{A}^{\mathrm{even}}(M, P \times_G \mathfrak{g})$. By what we said above, α gives rise to a morphism of commutative algebras, the *Weil homomorphism*

$$\mathbb{F}[\mathfrak{g}]^G \to \mathcal{A}^{\mathrm{even}}(M, \mathbb{F}), \ \ \Phi \mapsto \Phi(P, \alpha), \tag{A.30}$$

where $\Phi(P, \alpha) = [\tau, \Phi_\Lambda(\alpha_\tau)]$ with respect to a local section τ of P, compare (A.11). Moreover, if α is a sum of forms of strictly positive (and even) degrees, then the natural extension to $\mathbb{F}[[\mathfrak{g}]]^G$ is well defined.

A.31 Main Theorem. *Let $P \to M$ be a principal bundle with structure group G. Let D be a connection on P and R be the curvature of D. Then the Weil homomorphism*

$$\mathbb{F}[[\mathfrak{g}]]^G \to \mathcal{A}^{\mathrm{even}}(M, \mathbb{F}), \ \ \Phi \mapsto \Phi(P, R),$$

is a morphism of commutative algebras. Moreover,

1. *$\Phi(P, R)$ is closed and its cohomology class in $H^{\mathrm{even}}(M, \mathbb{F})$ does not depend on the choice of D;*

2. *if $P' \to M'$ is a G-principal bundle with connection D' and $F\colon P' \to P$ is a morphism over a smooth map $f\colon M' \to M$ with $D' = F^*D$, then $\Phi(P', R') = f^*\Phi(P, R)$.*

Proof. The first part is clear from the discussion leading to (A.30). Assertion (1) follows from Lemma A.19. Assertion (2) is immediate from the definitions. □

A.1 Chern Classes and Character. Let $G = \mathrm{Gl}(r, \mathbb{C})$ and $\mathfrak{g} = \mathfrak{gl}(r, \mathbb{C}) = \mathbb{C}^{r \times r}$. We get invariant homogeneous polynomial Φ_k of degree k on $\mathfrak{gl}(r, \mathbb{C})$ by setting

$$\det(tI + A) = \sum_{k=0}^{r} t^{r-k} \Phi_k(A). \tag{A.32}$$

If A is diagonalizable, then $\Phi_k(A)$ is the k-th elementary symmetric function of the eigenvalues of A. In particular, $\Phi_0 = 1$, $\Phi_1 = \mathrm{tr}$, and $\Phi_r = \det$. We note that $\Phi_k(A)$ is real if A is real.

Let $E \to M$ be a complex vector bundle of rank r and P be the bundle of frames of E. Then the structure group of P is $G = \mathrm{Gl}(r, \mathbb{C})$ and $P \times_G \mathfrak{g} = \mathrm{End}\, E$. Let D be a connection on E, or, what amounts to the same, on P. Then the k-th *Chern form* of E with respect to D is defined as

$$c_k(E, D) := \Phi_k\Big(P, \frac{i}{2\pi} R\Big), \tag{A.33}$$

where R is the curvature of D. The *total Chern form* of E with respect to D is

$$c(E, D) = \det\Big(I + \frac{i}{2\pi} R\Big) = 1 + c_1(E, D) + \cdots + c_r(E, D), \tag{A.34}$$

where r is the rank of E. The corresponding cohomology classes

$$\begin{aligned} c_k(E) &= [c_k(E, D)] \in H^{2k}(M, \mathbb{C}) \quad \text{and} \\ c(E) &= 1 + c_1(E) + \cdots + c_r(E) \in H^*(M, \mathbb{C}) \end{aligned} \tag{A.35}$$

are called the *Chern classes* and the *total Chern class* of E, respectively.

By definition, Chern classes of isomorphic bundles are equal. If E is trivial and D is the trivial connection on E, then $c_k(E, D) = 0$ for $k > 0$, and hence $c(E) = 1$. Therefore we view the Chern classes as a measure of the deviation of E from being trivial.

For a complex manifold M with complex structure J, we consider TM together with J as a complex vector bundle and use the shorthand $c_k(M)$ and $c(M)$ instead of $c_k(TM)$ and $c(TM)$, respectively.

A.36 Proposition. *If D is compatible with a Hermitian metric on E, then the Chern forms $c_k(E, D)$ are real. In particular, $c_k(E) \in H^{2k}(M, \mathbb{R}) \subset H^{2k}(M, \mathbb{C})$.*

Proof. With respect to a local orthonormal frame of E, that is, a local section of P, the principal part Ω of the curvature of D is skew-Hermitian, $\Omega = -\bar{\Omega}^t$. Hence

$$c(E, D) = \det\nolimits_\Lambda\Big(I + \frac{i}{2\pi}\Omega\Big) = \det\nolimits_\Lambda\Big(I - \frac{i}{2\pi}\bar{\Omega}\Big) = \bar{c}(E, D),$$

where we use that transposition does not change the determinant. □

A.37 Remark. There is a topological construction of Chern classes in $H^*(M,\mathbb{Z})$, see [MS]. The Chern classes we discuss here are their images under the natural map $H^*(M,\mathbb{Z}) \to H^*(M,\mathbb{R})$. In particular, the Chern classes here do not contain as much information as the ones coming from the topological construction. On the other hand, the construction here gives explicit differential forms representing the Chern classes, an advantage in some of the applications in differential geometry.

A.38 Proposition. *The Chern forms satisfy the following properties:*

1. *(Naturality)* $c(f^*E, f^*D) = f^*(c(E, D))$.
2. *(Additivity)* $c(E' \oplus E'', D' \oplus D'') = c(E', D') \wedge c(E'', D'')$.
3. $c_j(E^*, D^*) = (-1)^j c_j(E, D)$, *where* $E^* = \operatorname{Hom}(E, \mathbb{C})$ *is the dual bundle with induced connection* D^*.

The corresponding properties of Chern classes follow as a corollary.

Proof of Proposition A.38. The first assertion is immediate from $f^*\Omega_D = \Omega_{f^*D}$. With respect to local frames of E' and E'' over an open subset $U \subset M$, the principal part of the curvature of $D = D' \oplus D''$ is

$$\Omega = \begin{pmatrix} \Omega' & 0 \\ 0 & \Omega'' \end{pmatrix}.$$

Over U we have

$$\begin{aligned} c(E' \oplus E'', D' \oplus D'') &= \det{}_\wedge \left(\begin{pmatrix} I + \frac{i}{2\pi}\Omega' & 0 \\ 0 & I + \frac{i}{2\pi}\Omega'' \end{pmatrix} \right) \\ &= \det{}_\wedge \left(I + \frac{i}{2\pi}\Omega' \right) \wedge \det{}_\wedge \left(I + \frac{i}{2\pi}\Omega'' \right) \\ &= c(E', D') \wedge c(E'', D''), \end{aligned}$$

compare (A.27). This proves the second assertion. With respect to a local frame, $-\Omega^t$ is the principal part of the curvature of D^*. Hence the last assertion. □

A.39 Examples. 1) If the bundle $E \to M$ is trivial and D is the trivial connection on E, then $c(E, D) = 1$. In particular, $c(E) = 1$. More generally, if E has a flat connection, then $c(E) = 1$.

2) If E has rank r and splits as $E = E' \oplus E''$ with E'' trivial of rank s, then $c_j(E) = 0$ for $j > r - s$.

3) Let S be the oriented surface of genus g, endowed with a Riemannian metric and the corresponding Levi-Civita connection D. The rotation $J = J_p$ by a positive right angle in T_pS, $p \in S$, is a parallel field of complex structures which turns S into a Kähler manifold of complex dimension 1 and TS into a complex line bundle over S. Let $p \in S$ and (v, w) be an oriented orthonormal basis of T_pS. Then $w = Jv$ and $K(p) = -\langle R(v, w)v, w\rangle$, and hence

$$\Omega_p(v, w) = -i\,K(p),$$

where K is the Gauss curvature of S. Therefore

$$c_1(TS, D) = \frac{K}{2\pi}\, dA,$$

where dA denotes the (oriented) area form. The Gauss–Bonnet formula gives

$$\int_S c_1(TS, D) = \chi(S) = \text{Euler number of } S.$$

We make this more explicit in the case of $S = \mathbb{C}P^1 = S^2$. On the open subset $U = \{[z_0, z_1] \in S \mid z_1 \neq 0\}$ of S, consider the coordinates $x + iy = z := z_0/z_1$. Let $X = \partial/\partial x$ and $Y = \partial/\partial y$ be the corresponding coordinate vector fields. The complex structure on S turns TS into a complex line bundle. The standard metric of constant curvature 1 is the real part of the Hermitian metric on TS which, over U, is given by

$$h(X, X) = \frac{4}{(1 + |z|^2)^2}. \tag{A.40}$$

By Proposition 3.21.1, the connection form of the Chern connection D is

$$\omega = h^{-1}\partial h = (1 + |z|^2)^2 \partial\big((1 + |z|^2)^{-2}\big) = \frac{-2\bar{z}dz}{1 + |z|^2}.$$

By Proposition 3.21.4, the principal part of the curvature is

$$\Omega = \bar{\partial}\omega = \frac{2dz \wedge d\bar{z}}{(1 + |z|^2)^2}.$$

Hence the first Chern form

$$c_1(TS, D) = \frac{i\, dz \wedge d\bar{z}}{\pi(1 + |z|^2)^2} = \frac{2dx \wedge dy}{\pi(1 + |z|^2)^2}.$$

It follows that

$$\begin{aligned}
\int_S c_1(TS, D) &= \int_U c_1(TS, D) \\
&= \frac{2}{\pi}\int_0^\infty \int_0^{2\pi} \frac{\rho\, d\rho d\varphi}{(1 + \rho^2)^2} \\
&= 4\int_0^\infty \frac{\rho\, d\rho}{(1 + \rho^2)^2} = 2\int_1^\infty \frac{du}{u^2} = 2,
\end{aligned}$$

where we substitute $x + iy = \rho \exp(i\varphi)$ and $u = 1 + \rho^2$, respectively. As expected, the right hand side is the Euler number of $\mathbb{C}P^1 = S^2$.

4) In Example 3.24, we consider the canonical metric of the tautological bundle U over the complex Grassmannian $G_{k,n}$ and compute the principal part of the curvature

of the Chern connection D. In the case $U \to G_{1,2} = \mathbb{C}P^1$ and with respect to the coordinates z as in the previous example, we obtain

$$\Omega(0) = -dz \wedge d\bar{z}.$$

It follows that $\int_{\mathbb{C}P^1} c_1(U, D) = -1$ and that $c_1(\mathbb{C}P^1) = -2c_1(U)$.

5) Let $\pi : S^{2m+1} \to \mathbb{C}P^m$ be the natural projection, $x \in S^{2m+1}$, and $p = \pi(x)$. Then the differential of π at x identifies the orthogonal complement $p^\perp$ with the tangent space of $\mathbb{C}P^m$ at p. However, this isomorphism depends on the choice of p. The ambiguity is resolved by passing from $p^\perp$ to $\mathrm{Hom}(p, p^\perp)$. In other words, $T\mathbb{C}P^m \cong \mathrm{Hom}(U, U^\perp)$, where $U^\perp$ is the orthogonal complement of the tautological bundle $U \to \mathbb{C}P^m$ in the trivial bundle $\mathbb{C}^{m+1} \times CP^m \to \mathbb{C}P^m$ with respect to the standard Hermitian metric.

By the naturality of Chern classes, the first Chern class $c_1(U)$ has value -1 on the complex line $\mathbb{C}P^1 \subset \mathbb{C}P^m$. Now $\mathrm{Hom}(U, U) \cong \mathbb{C} \times \mathbb{C}P^m$, the trivial line bundle over $\mathbb{C}P^m$. Hence

$$T\mathbb{C}P^m \oplus \mathbb{C} = \mathrm{Hom}(U, U^\perp \oplus U) = \mathrm{Hom}(U, \mathbb{C}^{m+1}) \cong U^* \oplus \cdots \oplus U^*,$$

where we have $m+1$ summands on the right hand side. Therefore

$$c(\mathbb{C}P^m) = c(T\mathbb{C}P^m \oplus \mathbb{C}) = (1 - c_1(U))^{m+1} = (1+a)^{m+1},$$

where a is the generator of $H^2(\mathbb{C}P^m, \mathbb{C})$ which has value 1 on the complex line $\mathbb{C}P^1 \subset \mathbb{C}P^m$. In particular,

$$c_1(\mathbb{C}P^m) = (m+1)a. \tag{A.41}$$

It follows that $c_m(\mathbb{C}P^m)[\mathbb{C}P^m] = m+1$, the Euler number of $\mathbb{C}P^m$.

A.42 Remark. If E is a complex vector bundle of rank r over a manifold M, then $c_r(E)$ is the Euler class of E, considered as an oriented real vector bundle of rank $2r$, see Proposition A.59 below.

On $\mathfrak{gl}(r, \mathbb{C}) = \mathbb{C}^{r\times r}$, consider the invariant formal power series

$$\Phi(A) := \operatorname{tr} \exp A = r + \operatorname{tr} A + \frac{1}{2} \operatorname{tr} A^2 + \cdots. \tag{A.43}$$

Let $E \to M$ be a complex vector bundle of rank r and P be the bundle of frames of E. Let D be a connection on E and R be the curvature of D. Then the cocycle

$$\mathrm{ch}(E, D) := \Phi\Big(P, \frac{i}{2\pi} R\Big) \in \mathcal{A}^{\mathrm{even}}(M, \mathbb{C}) \tag{A.44}$$

is called the *Chern character* of E with respect to D. Its cohomology class $\Phi(E)$ in $H^{\mathrm{even}}(M, \mathbb{C})$ is called the Chern character of E. If A is diagonal, $A = D(a_1, \dots, a_r)$, then

$$\begin{aligned} \operatorname{tr} A &= a_1 + \cdots + a_r = \sigma_1(a_1, \dots, a_r), \\ \operatorname{tr} A^2 &= a_1^2 + \cdots a_r^2 = \sigma_1(a_1, \dots, a_r)^2 - 2\sigma_1(a_1, \dots, a_r), \end{aligned}$$

where $\sigma_1, \sigma_2, \dots$ denotes the sequence of elementary symmetric functions. Since diagonal matrices are dense in $\mathbb{C}^{r\times r}$, we conclude that

$$\operatorname{tr} A = \Phi_1(A), \quad \text{and} \quad \frac{1}{2}\operatorname{tr} A^2 = \frac{1}{2}\Phi_1(A)^2 - \Phi_2(A), \tag{A.45}$$

where Φ_1 and Φ_2 are as in (A.32). By the fundamental theorem on symmetric polynomials, the symmetric function $\operatorname{tr} D(a_1, \dots, a_r)^k$ can be expressed uniquely as a polynomial in $\sigma_1, \dots, \sigma_r$. It follows that $\operatorname{ch}(E, D)$ can be expressed as a polynomial in the Chern forms $c_k(E, D)$. From what we noted above, we have

$$\operatorname{ch}(E, D) = r + c_1(E, D) + \Big(\frac{1}{2}c_1(E, D)^2 - c_2(E, D)\Big) + \cdots. \tag{A.46}$$

We leave it as an exercise to compute the term of degree 6 in this formula.

A.47 Exercise. Follow the argument in Proposition A.38.2 and show that

$$\operatorname{ch}(E' \oplus E'', D' \oplus D'') = \operatorname{ch}(E', D') + \operatorname{ch}(E'', D'').$$

Recall that the derived action of the Lie algebra $\mathfrak{gl}(r, \mathbb{C}) \times \mathfrak{gl}(s, \mathbb{C})$ on $\mathbb{C}^r \otimes \mathbb{C}^s$ is given by $(A, B)(x, y) = (Ax) \otimes y + x \otimes (By)$ and show that

$$\operatorname{ch}(E' \otimes E'', D' \otimes D'') = \operatorname{ch}(E', D') \wedge \operatorname{ch}(E'', D'').$$

It follows that the Chern character induces a homomorphism from the Grothendieck ring $K(M)$ generated by equivalence classes of complex vector bundles over M to the even cohomology ring $H^{\text{even}}(M, \mathbb{C})$.

A.2 Euler Class. On $\mathfrak{gl}(2r, \mathbb{R}) = \mathbb{R}^{2r\times 2r}$, consider the homogeneous polynomial

$$\operatorname{Pf}(A) := \frac{1}{2^r r!}\sum_\sigma \varepsilon(\sigma) a_{\sigma(1)\sigma(2)} \dots a_{\sigma(2r-1)\sigma(2r)} \tag{A.48}$$

of degree r, the *Pfaffian* of A. It is easy to show that

$$\operatorname{Pf}(BAB^t) = \operatorname{Pf}(A) \det B. \tag{A.49}$$

Therefore the Pfaffian defines an invariant polynomial on $\mathfrak{so}(2r)$, the Lie algebra of the group of rotations $\mathrm{SO}(2r)$. The normal form on $\mathfrak{so}(2r)$ is

$$A = \begin{pmatrix} A_1 & & \\ & \ddots & \\ & & A_r \end{pmatrix} \quad \text{with} \quad A_j = \begin{pmatrix} 0 & -a_j \\ a_j & 0 \end{pmatrix}. \tag{A.50}$$

For such a matrix A, we have $\operatorname{Pf}(A) = (-1)^r a_1 \dots a_r$. By invariance, we conclude that on $\mathfrak{so}(2r)$,

$$\operatorname{Pf}(A)^2 = \det A. \tag{A.51}$$

Let $E \to M$ be an oriented real vector bundle of rank $2r$ with a Riemannian metric g. Then the principal bundle P of oriented orthonormal frames of E has structure group $\mathrm{SO}(2r)$. If D is a connection on P, that is, a metric connection on E, and R denotes the curvature of D, then the associated form

$$\chi(E, g, D) := \mathrm{Pf}\left(P, \frac{1}{2\pi} R\right) \tag{A.52}$$

is called the *Euler form* of E with respect to g and D. It is clear that

$$\chi(f^*E, f^*g, f^*D) = f^*\chi(E, g, D) \tag{A.53}$$

and that

$$\chi(E \oplus E', g \oplus g', D \oplus D') = \chi(E, g, D) \wedge \chi(E', g', D'). \tag{A.54}$$

Let g' be another Riemannian metric on E and write $g' = g(B\cdot, \cdot)$ where B is a field of symmetric and positive definite endomorphisms of E. Let $C = \sqrt{B}$, then $g'(x, y) = g(Cx, Cy)$, and hence C is an isomorphism of E with $C^*g = g'$. It follows that $\chi(E, g', C^*D) = \chi(E, g, D)$. We conclude that the cohomology class $\chi(E)$ of $\chi(E, g, D)$ in $H^{2r}(M, \mathbb{R})$ does not depend on g and D. It is called the *Euler class* of E.

A.55 Remark. If E has a Riemannian metric g with a flat metric connection D, then $\chi(E, g, D) = 0$ and hence also $\chi(E) = 0$. However, there are vector bundles E with flat non-metric connections such that $\chi(E) \neq 0$, see for example [MS, 312 pp]. In fact, in the definition of the Euler form we need a Riemannian metric and a corresponding metric connection on E.

Our next aim is to compare Euler class and Chern classes. We identify $\mathbb{C}^r$ with $\mathbb{R}^{2r}$ via the correspondence

$$(z^1, \dots, z^r) \longleftrightarrow (x^1, y^1, \dots, x^r, y^r),$$

where $z^j = x^j + iy^j$. This induces an inclusion $\mathrm{U}(r) \hookrightarrow \mathrm{SO}(2r)$ of the unitary group and a corresponding inclusion $\mathfrak{u}(r) \hookrightarrow \mathfrak{so}(2r)$, $A \mapsto A_{\mathbb{R}}$, of Lie algebras. Under this inclusion, diagonal matrices

$$A = \begin{pmatrix} ia_1 & & \\ & \ddots & \\ & & ia_r \end{pmatrix} \mapsto \begin{pmatrix} A_1 & & \\ & \ddots & \\ & & A_r \end{pmatrix} = A_{\mathbb{R}}, \tag{A.56}$$

where A_j is as above. For such a diagonal A, $\det(iA) = (-1)^r a_1 \dots a_r = \mathrm{Pf}(A_{\mathbb{R}})$. By invariance,

$$\det(iA) = \mathrm{Pf}(A_{\mathbb{R}}) \tag{A.57}$$

for all $A \in \mathfrak{u}(r)$.

Suppose now that $E \to M$ is a complex vector bundle of rank r. Endow E with a Hermitian metric h and a corresponding Hermitian connection D. If we consider E

as an oriented real vector bundle of rank $2r$, the real part $g = \operatorname{Re} h$ of the Hermitian metric defines a Riemannian metric on E and D is metric with respect to g.

Let ω and Ω be the connection and curvature forms of D with respect to a local unitary frame of E. If we consider E as an oriented real vector bundle and choose D as a metric connection for the Riemannian metric g as above, then $\omega_{\mathbb{R}}$ and $\Omega_{\mathbb{R}}$ are the connection and curvature forms of D with respect to the induced oriented and orthonormal frame of E. By (A.33) and (A.57),

$$c_r(E, D) = \det\Big(\frac{i}{2\pi}\Omega\Big) = \operatorname{Pf}\Big(\frac{1}{2\pi}\Omega_{\mathbb{R}}\Big) = \chi(E, g, D). \tag{A.58}$$

A.59 Proposition. *For a complex vector bundle $E \to M$ of rank r, $\chi(E) = c_r(E)$.* □

Proposition A.59 and Equations A.53, A.54 imply that the Euler class of (the tangent bundle of) an oriented closed manifold M, applied to the fundamental class of M, gives the Euler number of M. This is the celebrated Chern–Gauß–Bonnet formula.

Appendix B Symmetric Spaces

In this appendix we collect a few of the basic facts about symmetric spaces. Good references for unproved statements and further reading are [Bor], [Hel], [Wo].

We say that a Riemannian manifold M is a *symmetric space* if M is connected and if, for all $p \in M$, there is an isometry s_p of M with

$$s_p(p) = p \quad \text{and} \quad ds_p(p) = -\operatorname{id}. \tag{B.1}$$

We then call s_p the *geodesic reflection* at p.

B.2 Exercise. Let M be a connected Riemannian manifold and f_1, f_2 be isometries of M. Suppose there is a point $p \in M$ with $f_1(p) = f_2(p)$ and $df_1(p) = df_2(p)$. Then $f_1 = f_2$.

We say that a Riemannian manifold M is *locally symmetric* if each point $p \in M$ has a neighborhood U in M with an isometry $s_p \colon U \to U$ such that $s_p(p) = p$ and $ds_p(p) = -\operatorname{id}$.

B.3 Exercises. 1) Let M be a symmetric space and $\tilde{M} \to M$ be a covering. Show that $\tilde{M}$ with the induced Riemannian metric is also a symmetric space.

2) Let $M = M_1 \times M_2$ be a Riemannian product. Show that M is symmetric if and only if M_1 and M_2 are symmetric.

3) Prove the corresponding assertions in the case of locally symmetric spaces.

B.4 Proposition. *Let M be a locally symmetric space and F be a tensor field on M of type (k,l). If F is invariant under the local isometries s_p, $p \in M$, then $F = 0$ if $k + l$ is odd.*

For example, if M is locally symmetric, then the curvature tensor is parallel, $\nabla R = 0$. In fact, it is not hard to see that M is locally symmetric if and only if the curvature tensor of M is parallel. Another application of Proposition B.4: If M is locally symmetric and J is an almost complex structure on M which is invariant under the local isometries s_p, $p \in M$, then J is parallel.

Proof of Proposition B.4. Let $p \in M$, $v_1, \dots, v_k \in T_pM$, and $\varphi_1, \dots, \varphi_l \in T_p^*M$. Since $s_p(p) = p$, $ds_p(p) = -\operatorname{id}$, and $(s_p^*F)_p = F_p$, we conclude

$$\begin{aligned} F_p(v_1, \dots, v_k, \varphi_1, \dots, \varphi_l) &= F_p(s_{p*}v_1, \dots, s_{p*}v_k, s_p^*\varphi_1, \dots, s_p^*\varphi_l) \\ &= F_p(-v_1, \dots, -v_k, -\varphi_1, \dots, -\varphi_l) \\ &= (-1)^{k+l} F_p(v_1, \dots, v_k, \varphi_1, \dots, \varphi_l). \end{aligned}$$ □

The fundamental theorem about locally symmetric spaces is as follows:

B.5 Theorem (Cartan). *Let M and N be locally symmetric Riemannian spaces, where M is simply connected and N is complete. Let $p \in M$ and $q \in N$, and let $A \colon T_pM \to T_qN$ be a linear isomorphism preserving inner products and curvature tensors, $A^*g_q^N = g_p^M$ and $A^*R_q^N = R_p^M$. Then there is a unique local isometry $F \colon M \to N$ with $F(p) = q$ and $F'(p) = A$.*

As for a proof, see for example the more general Theorems 1.1.36 in [CEb] or III.5.1 in [Sa][23].

B.6 Corollary. *Let M be a complete and simply connected locally symmetric space. Then M is a symmetric space. More precisely, the differentials of the isometries of M fixing a point p in M are given precisely by the orthogonal transformations of T_pM preserving the curvature tensor R_p^M.* □

This is one of the most important immediate applications of Theorem B.5. Another immediate application asserts that a simply connected symmetric space is determined, up to isometry, by the inner product and the curvature tensor at a point (using Propositions B.7 and B.8 below). This implies, for example, the uniqueness of the model Riemannian manifolds of constant sectional curvature.

B.7 Proposition. *If M is symmetric, then M is homogeneous.*

Proof. Let $c\colon [a,b] \to M$ be a geodesic in M, and let m be the midpoint of c. Then the geodesic reflection s_m of M at m reflects c about m, hence $s_m(c(a)) = c(b)$.

Since M is connected, any two points $p, q \in M$ can be connected by a piecewise geodesic c. If $c\colon [a,b] \to M$ is such a curve, then there is a subdivision $a = t_0 < t_1 < \dots < t_k = b$ of $[a,b]$ such that $c|[t_{i-1}, t_i]$ is a geodesic, $1 \le i \le k$. By the first part of the proof, there is an isometry f_i of M mapping $c(t_{i-1})$ to $c(t_i)$. Hence $f_k \circ \dots \circ f_1$ is an isometry of M mapping $p = c(a)$ to $q = c(b)$. □

B.8 Proposition. *If M is homogeneous, then M is complete.*

Proof. Let $p \in M$. Then there is an $\varepsilon > 0$ such that the closed ball $\bar B(p,\varepsilon)$ of radius ε about p is compact. Since M is homogeneous, the closed ball $\bar B(q,\varepsilon)$ of radius ε about any point $q \in M$ is compact. It follows that M is complete. □

Let M be a symmetric space. Let $c\colon [a,b] \to M$ be a piecewise smooth curve. An isometry $f\colon M \to M$ is called *transvection* along c if $f(c(a)) = c(b)$ and if $df(c(a))$ is equal to parallel translation along c from $c(a)$ to $c(b)$.

Let p be a point in M. Let $c\colon \mathbb{R} \to M$ be a geodesic through p. Let s_t be the geodesic reflection about $c(t)$ and $f_t = s_{t/2} \circ s_0$. Then f_t is an isometry of M with $f_t(c(\tau)) = c(\tau + t)$ for all $\tau \in \mathbb{R}$.

Let X be a parallel vector field along c. Then $ds_0 \circ X$ is parallel along $s_0 \circ c$. Since $s_0(c(s)) = c(-s)$ and $ds_0(p) = -\operatorname{id}$, we have

$$ds_0 X(s) = -X(-s).$$

Similarly,

$$ds_{t/2} X(s) = -X(t-s).$$

We conclude that, for all $\tau \in \mathbb{R}$,

$$df_t X(\tau) = X(\tau + t).$$

[23]Both references contain also introductions to symmetric spaces.

Hence $df_t(c(\tau))$ is parallel translation along c from $c(\tau)$ to $c(\tau + t)$. Hence f_t is a transvection along c; more precisely, it is the transvection along $c|[\tau, \tau + t]$ for any $\tau \in \mathbb{R}$. It follows that the family (f_t) is a smooth one-parameter group of isometries of M which shifts c and whose derivatives correspond to parallel translation along c. We call it the *one-parameter group of transvections* along c. Associated to this family we have the Killing field

$$X(p) = \partial_t(f_t(p))|_{t=0}.$$

We call X the *infinitesimal transvection* along c.

B.9 Lemma. *Let $c\colon \mathbb{R} \to M$ be a geodesic with $c(0) = p$. Then a Killing field X on M is the infinitesimal transvection along c if and only if $X(p) = c'(0)$ and $\nabla X(p) = 0$.*

Proof. Let X be the infinitesimal transvection along c. Then $X(p) = c'(0)$ since $f_t(p) = c(t)$. As for the covariant derivative of X in p, let $v \in T_pM$ and $\sigma = \sigma(\tau)$ be a smooth curve with $\sigma(0) = p$ and $\sigma'(0) = v$. Then

$$\begin{aligned}\nabla_v X &= \nabla_\tau \partial_t(f_t(\sigma(\tau)))|_{t=\tau=0} \\ &= \nabla_t \partial_\tau(f_t(\sigma(\tau)))|_{\tau=t=0} \\ &= \nabla_t(df_t v)|_{t=0} = 0\end{aligned}$$

since df_t is parallel translation along c. Since M is connected, a Killing field on M is determined by its value and covariant derivative at one point. □

B.10 Proposition. *Let M be a symmetric space, p be a point in M. Let $c\colon [a, b] \to M$ be a geodesic loop at p. Then c is a closed geodesic, $c'(a) = c'(b)$.*

Proof. Let $v \in T_pM$ and X be the infinitesimal transvection along the geodesic through p with initial velocity v. Then $X(p) = v$ and $\nabla X(p) = 0$, by Lemma B.9. Since X is a Killing field, $X \circ c$ is a Jacobi field along c, see Exercise 1.5. Recall that for any two Jacobi fields V, W along c, the function $\langle V', W\rangle - \langle V, W'\rangle$ is constant, where the prime indicates covariant derivative along c. We apply this to the Jacobi fields $X \circ c$ and tc',

$$\langle (X \circ c)', tc'\rangle - \langle X \circ c, c'\rangle = \text{const.}$$

Since $c(a) = c(b) = p$ and $\nabla X(p) = 0$, we get

$$\langle X(p), c'(a)\rangle = \langle X(p), c'(b)\rangle.$$

Since $X(p) = v \in T_pM$ was arbitrary, we conclude that $c'(a) = c'(b)$. □

B.11 Corollary. *The fundamental group of a symmetric space is Abelian.*

Proof. Let M be a symmetric space and p be a point in M. Shortest loops in homotopy classes of loops at p are geodesic and hence closed geodesics through p, by Proposition B.10. The geodesic reflection at p maps closed geodesics through p into their inverses. Hence s_p induces the inversion on $\pi_1(M, p)$. Now a group is Abelian if its inversion is a homomorphism. □

B.12 Definition. Let M be a symmetric space. We say that M is of *compact type* or, respectively, *non-compact type* if the Ricci curvature of M is positive or, respectively, negative. We say that M is of *Euclidean type* if M is flat, that is, if the sectional curvature of M vanishes.

If M is a symmetric space of compact type, then M is compact and the fundamental group of M is finite, by the theorem of Bonnet–Myers. Flat tori and their Riemannian covering spaces are symmetric spaces of Euclidean type.

B.13 Proposition. *If M is a symmetric space of non-compact type, then M does not have non-trivial closed geodesics.*

Proof. Let $c\colon \mathbb{R} \to M$ be a closed geodesic of unit speed. Set $p = c(0)$ and $u = c'(0)$. Since M is of non-compact type, $\operatorname{Ric}(u, u) < 0$. Hence the symmetric endomorphism $R(\cdot, u)u$ of T_pM has a negative eigenvalue κ and a corresponding unit eigenvector v perpendicular to u. Let V be the parallel vector field along c with $V(0) = v$. Since R is parallel, $R(V, c')c' = \kappa V$ along c. Therefore $J = \cosh(\sqrt{-\kappa}t)V$ is the Jacobi field along c with $J(0) = v$ and $J'(0) = 0$.

Let X be the infinitesimal transvection through p with $X(p) = v$, that is, along the geodesic determined by v. Then $Y = X \circ c$ is a Jacobi field along c with the same initial conditions as J: $Y(0) = v$ and $Y'(0) = 0$, hence $Y = J$. Now as the restriction of a vector field on M to c, Y is periodic along c. On the other hand, J is definitely not periodic. Contradiction. □

B.14 Corollary. *Let M be a symmetric space of non-compact type. Then:*

1) *M is simply connected.*

2) *If G denotes the component of the identity of the group of isometries of M and $K \subset G$ the stabilizer of a point $p \in M$, then K is connected.*

Proof. Since shortest loops in homotopy classes of loops at p are geodesic loops, the first assertion is immediate from Propositions B.10 and B.13. As for the second, we note that the last piece of the long exact homotopy sequence of the fibration $G \to M$, $g \mapsto gp$, reads

$$1 = \pi_1(M) \to \pi_0(K) \to \pi_0(G) = 1. \qquad \square$$

B.15 Remark. In Remark B.29 we show that a symmetric space has non-negative sectional curvature iff its Ricci curvature is non-negative and, similarly, that it has non-positive sectional curvature iff its Ricci curvature is non-positive. In particular, a symmetric space is of Euclidean type if and only if its Ricci curvature vanishes. Furthermore, if M is a symmetric space of non-compact type, then M is diffeomorphic to $\mathbb{R}^n$, $n = \dim M$, by the theorem of Hadamard–Cartan and Corollary B.14.1. It follows that, for K and G as in Corollary B.14.2, K is a deformation retract of G. This improves Corollary B.14.2.

B.1 Symmetric Pairs. Let (G, K) be a pair consisting of a Lie group G and a closed subgroup K. We say that (G, K) is a *symmetric pair* if $M = G/K$ is connected and if there is an involutive automorphism $\sigma : G \to G$ with

$$F_0 \subset K \subset F, \tag{B.16}$$

where $F = \{g \in G \mid \sigma(g) = g\}$ and F_0 denotes the component of the identity of F. In what follows, (G, K) is a symmetric pair with involutive automorphism σ.

Let $\mathfrak{g}$ be the Lie algebra of left-invariant vector fields of G, via evaluation identified with the tangent space T_eG of G at the neutral element $e \in G$. Denote by σ_* the differential of σ at e, and let

$$\mathfrak{k} = \{X \in \mathfrak{g} \mid \sigma_* X = X\}, \quad \mathfrak{p} = \{X \in \mathfrak{g} \mid \sigma_* X = -X\}, \tag{B.17}$$

the eigenspaces of σ_* for the eigenvalues 1 and -1, respectively. Since σ_* is involutive,

$$\mathfrak{g} = \mathfrak{k} + \mathfrak{p}. \tag{B.18}$$

Since $F_0 \subset K \subset F$, $\mathfrak{k}$ is the Lie algebra of K. Since $K \subset F$, $\sigma(kgk^{-1}) = k\sigma(g)k^{-1}$ for all $k \in K$ and all $g \in G$, and hence σ_* commutes with all Ad_k, $k \in K$. Therefore

$$\mathrm{Ad}_k(\mathfrak{k}) \subset \mathfrak{k} \quad \text{and} \quad \mathrm{Ad}_k(\mathfrak{p}) \subset \mathfrak{p} \quad \text{for all } k \in K. \tag{B.19}$$

Furthermore,

$$[\mathfrak{k}, \mathfrak{k}] \subset \mathfrak{k}, \quad [\mathfrak{k}, \mathfrak{p}] \subset \mathfrak{p}, \quad [\mathfrak{p}, \mathfrak{p}] \subset \mathfrak{k}. \tag{B.20}$$

In particular, $\mathfrak{k}$ is perpendicular to $\mathfrak{p}$ with respect to the Killing form of $\mathfrak{g}$. As for the proof of (B.20), the first and second inclusion follow from (B.19) above. For the proof of the third, let $X, Y \in \mathfrak{p}$. Since σ_* is an automorphism of $\mathfrak{g}$, we have

$$\sigma_*[X, Y] = [\sigma_* X, \sigma_* Y] = [-X, -Y] = [X, Y]$$

and hence $[X, Y] \in \mathfrak{k}$. Note that the first two inclusions also follow by a similar argument.

Since we divide by K on the right, G acts from the left on M. We denote by $gp \in M$ the image of $p \in M$ under left multiplication by $g \in G$. Sometimes it will be convenient to distinguish between the element $g \in G$ and the diffeomorphism of M given by left multiplication with g. Then we will use λ_g to denote the latter, $\lambda_g(p) := gp$.

B.21 Exercise. Let M be a manifold and G be a Lie group which acts on M. For $X \in \mathfrak{g}$, define a smooth vector field X^* on M by

$$X^*(p) = \partial_t(e^{tX}(p))|_{t=0}.$$

By definition, $\exp(tX)$ is the flow of X^*. Show that $[X, Y]^* = -[X^*, Y^*]$.

We denote by $o = [K]$ the distinguished point of M and by $\pi : G \to M$ the canonical map, $\pi(g) = [gK] = \lambda_g(o)$. We use $\pi_* = d\pi(e) \colon \mathfrak{p} \to T_oM$ to identify $\mathfrak{p}$ and T_oM. An easy computation shows that for all $k \in K$

$$\pi_* \circ \mathrm{Ad}_k = d\lambda_k(o) \circ \pi_*. \tag{B.22}$$

Thus with respect to the identification $\mathfrak{p} \cong T_oM$ via π_*, the isotropy representation of K on T_oM corresponds to the restriction of the adjoint representation of K to $\mathfrak{p}$.

We say that a symmetric pair is *Riemannian* if there is an inner product $\langle\cdot,\cdot\rangle$ on $\mathfrak{p}$ which is invariant under Ad_K, that is, under all Ad_k, $k \in K$. Note that such inner products are in one-to-one correspondence with G-invariant Riemannian metrics on M.

B.23 Example. Let M be a symmetric space and $o \in M$ be a preferred origin. Let G be the group $\mathrm{Iso}(M)$ of all isometries of M or the component of the identity in $\mathrm{Iso}(M)$. Let K be the stabilizer of o in G. Then (G, K) is a Riemannian symmetric pair with respect to the involution σ of G given by conjugation with the geodesic reflection s_o of M in o. The pull back of the inner product in T_oM to $\mathfrak{p}$ is invariant under Ad_K and turns (G, K) into a Riemannian symmetric pair with $G/K = M$.

B.24 Theorem. *Let (G, K) be a Riemannian symmetric pair with corresponding involution σ of G and inner product $\langle\cdot,\cdot\rangle$ on $\mathfrak{p}$. Let $M = G/K$ and endow M with the G-invariant Riemannian metric corresponding to $\langle\cdot,\cdot\rangle$. Then we have:*

1. *M is a symmetric space. The geodesic symmetry s at o is $s([gK]) = [\sigma(g)\cdot K]$. In particular, $s \circ \lambda_g = \lambda_{\sigma(g)} \circ s$.*

2. *For $X \in \mathfrak{p}$, the curve $e^{tX}(o)$, $t \in \mathbb{R}$, is the geodesic through o with initial velocity π_*X, and left multiplication by e^{tX}, $t \in \mathbb{R}$, is the one-parameter group of transvections along this geodesic.*

3. *With respect to the identification $\mathfrak{p} \cong T_oM$, Ad_K-invariant tensors on $\mathfrak{p}$ correspond to G-invariant tensor fields on M and these are parallel.*

4. *With respect to the identification $\mathfrak{p} \cong T_oM$, the curvature tensor R and Ricci tensor* Ric *of M at o are given by*

$$R(X, Y)Z = -[[X, Y], Z] = -[[X^*, Y^*], Z^*](p),$$
$$\mathrm{Ric}(X, Y) = -\frac{1}{2}B(X, Y),$$

where B denotes the Killing form of $\mathfrak{g}$ and X^*, Y^*, Z^* are associated to X, Y, Z as in Exercise B.21.

5. *With respect to the identification $\mathfrak{p} \cong T_oM$, a subspace $\mathfrak{q} \subset \mathfrak{p}$ is tangent to a totally geodesic submanifold of M through o if and only if $[[\mathfrak{q}, \mathfrak{q}], \mathfrak{q}] \subset \mathfrak{q}$. If the latter inclusion holds, then $N = \exp(\mathfrak{q})(o)$ is such a submanifold and is a symmetric space in the induced Riemannian metric.*

B.25 Remarks. 1) We can rewrite the first formula in B.24.4 as follows:

$$R(X, Y) = -\operatorname{ad}_{[X,Y]} \quad \text{or} \quad R(\cdot, Y)Z = -\operatorname{ad}_Z \circ \operatorname{ad}_Y,$$

where $X, Y, Z \in \mathfrak{p}$ and both maps are considered on $\mathfrak{p}$.

2) The totally geodesic submanifold in B.24.5 need not be closed. A good example for this is a non-rational line through 0 in a flat torus.

Proof of Theorem B.24. It is easy to see that $s([gK]) := [\sigma(g)\cdot K]$ defines an involutive smooth map s of M with $s \circ \lambda_g = \lambda_{\sigma(g)} \circ s$. Since s is involutive, s is a diffeomorphism of M. Moreover $s(o) = o$ and $ds(o) = -\operatorname{id}$.

Let $p = g(o) \in M$ and $u \in T_pM$. Choose $v \in T_oM$ with $d\lambda_g(v) = u$. By the G-invariance of the metric, we have $\|u\| = \|d\lambda_g(v)\| = \|v\|$ and hence

$$\|ds(u)\| = \|ds(d\lambda_g(v))\| = \|d\lambda_{\sigma(g)}(ds(v))\| = \|-d\lambda_{\sigma(g)}(v)\| = \|v\| = \|u\|.$$

It follows that s is an isometry. This completes the proof of (1).

Let $X \in \mathfrak{p}$ and X^* be the corresponding Killing field of M as in Exercise B.21. Since $X \in \mathfrak{p}$, we have

$$e^{tX}(s(p)) = s(e^{-tX}(p))$$

for all $p \in M$, hence $s_*X^* = -X^*$. Let $u \in T_oM$. Since s is an isometry, we get

$$-\nabla_u X^* = ds(\nabla_u X^*) = \nabla_{ds(u)}(s_*X^*) = \nabla_u X^*,$$

and hence $\nabla X^*(o) = 0$. Hence X^* is the infinitesimal transvection with $X^*(o) = \pi_*X$. This proves (2).

It is clear that Ad_K-invariant tensors on $\mathfrak{p}$ correspond to G-invariant tensor fields on M. Now the one-parameter groups $(\exp(tX))$, $X \in \mathfrak{p}$, are the transvections along the corresponding geodesics through o. It follows that G-invariant tensor fields are parallel at o and hence parallel everywhere in M. Hence (3).

Let $X, Y, Z \in \mathfrak{p}$ and X^*, Y^*, Z^* be the corresponding Killing fields of M as in Exercise B.21. Then X^* Y^*, and Z^* are parallel at o. Let $u = Z^*(o) = \pi_*Z$ and compute $\nabla_u[X^*, Y^*]$ using the differential equation for Killing fields. This gives the first equation in (4). In the proof of the second we can assume that $X = Y$, by the symmetry of the Ricci tensor. Since $X \in \mathfrak{p}$, we have $\operatorname{ad} X(\mathfrak{k}) \subset \mathfrak{p}$ and $\operatorname{ad} X(\mathfrak{p}) \subset \mathfrak{k}$. Therefore $\operatorname{tr} \operatorname{ad}^2 X|\mathfrak{k} = \operatorname{tr} \operatorname{ad}^2 X|\mathfrak{p}$, and the claim about the Ricci tensor follows.

It remains to prove (5). Note that for a totally geodesic submanifold N through the origin o, $R(u, v)w \in T_oN$ for all $u, v, w \in T_oN$. Hence the necessity of the condition on $\mathfrak{q}$ is immediate from the formula for R in (3).

Suppose now that $[[\mathfrak{q}, \mathfrak{q}], \mathfrak{q}] \subset \mathfrak{q}$. Let $\mathfrak{l} = [\mathfrak{q}, \mathfrak{q}]$. Then $\mathfrak{l}$ is a Lie subalgebra of $\mathfrak{k}$ and $\mathfrak{h} = \mathfrak{l} + \mathfrak{q}$ is a Lie subalgebra of $\mathfrak{g}$. Let H be the corresponding connected Lie subgroup of G. Then the orbit N of o under H, $N = H(o)$, is a submanifold of M. Now N is totally geodesic at o since $N = \exp(\mathfrak{q})(o)$ locally about o. Since H is transitive on N and H acts isometrically on M, N is totally geodesic everywhere. Since M is a complete Riemannian manifold and N is connected we have $N = \exp(\mathfrak{q})(o)$. In particular, N is complete. Since N is invariant under s, $s|N$ is the geodesic symmetry of N in o. Since N is homogeneous, N is symmetric. □

B.26 Remark (Coverings). The groups K with $F_0 \subset K \subset F$ correspond to Riemannian coverings

$$G/F_0 \to G/K \to G/F.$$

For example, if $M = S^n \subset \mathbb{R}^{n+1}$, the unit sphere of dimension n, then the component G of the identity in $\mathrm{Iso}(M)$ is equal to $\mathrm{SO}(n+1)$. If we let s be the reflection in a chosen unit vector $v \in S^n$ and σ be conjugation with s, then the stabilizer K of v is isomorphic to $\mathrm{SO}(n)$ and $K = F_0 \neq F$. If we pass to the quotient $\mathbb{R}P^n$ keeping G, then we have $K = F$. A similar phenomenon occurs for some of the other symmetric spaces of compact type.

It is remarkable that the fixed point set of an involution of a compact and simply connected Lie group is connected, see [Hel, Theorem VII.7.2]. Thus the symmetric space associated to a Riemannian symmetric pair (G, K) with G compact and simply connected is simply connected. For example, the unit sphere S^n is associated to the Riemannian symmetric pair $(\mathrm{Spin}(n+1), \mathrm{Spin}(n))$. Real projective space $\mathbb{R}P^n$ is a homogeneous space of $\mathrm{Spin}(n+1)$, but is not associated to a Riemannian symmetric pair (G, K) with G simply connected.

B.27 Remark (Effective Pairs). Let (G, K) be a Riemannain symmetric pair and $N \subset G$ be the normal subgroup of elements acting trivially on $M = G/K$,

$$N = \{g \in G \mid g \text{ acts as identity on } M\}.$$

Then $N \subset K$, hence the involution σ of G factors over G/N. Since M is connected, an isometry of $M = G/K$ fixing the preferred origin o of M is determined by its differential at o, see Exercise B.2. Hence

$$N = \{k \in K \mid \mathrm{Ad}_k \text{ is the identity on } \mathfrak{p}\}.$$

In particular, the adjoint action of K on $\mathfrak{p}$ factors over K/N. It follows that $(G/N, K/N)$ is a Riemannian symmetric pair with $(G/N)/(K/N) = G/K = M$ and Lie algebra $\mathfrak{g}/\mathfrak{n} \cong \mathfrak{k}/\mathfrak{n} + \mathfrak{p}$.

We say that the Riemannian symmetric pair (G, K) is *effective* or, respectively, *infinitesimally effective* if N is trivial or, respectively, a discrete subgroup of G. The pair (G, K) is effective iff G is a subgroup of the isometry group of M. The above pair $(G/N, K/N)$ is *the* effective pair associated to (G, K). Thus it becomes plausible that, for many purposes, it is sufficient to consider effective or infinitesimally effective pairs.

B.28 Remark (Killing Form). Let (G, K) be an infinitesimally effective Riemannian symmetric pair with K/N compact, where N is as in Remark B.27. Let $X \in \mathfrak{k}$. From (B.20) we know that $\mathrm{ad}_X(\mathfrak{k}) \subset \mathfrak{k}$ and $\mathrm{ad}_X(\mathfrak{p}) \subset \mathfrak{p}$. Hence

$$B(X, X) = \mathrm{tr}(\mathrm{ad}_X|\mathfrak{k})^2 + \mathrm{tr}(\mathrm{ad}_X|\mathfrak{p})^2 = B_K(X, X) + \mathrm{tr}(\mathrm{ad}_X|\mathfrak{p})^2,$$

where B is the Killing form of G and B_K the Killing form of K. Since $N \subset K$ is a discrete subgroup and K/N is compact, $B_K \leq 0$ and hence $B(X, X) \leq \mathrm{tr}(\mathrm{ad}_X|\mathfrak{p})^2$.

On the other hand, since N is discrete and $X \in \mathfrak{k}$, $\mathrm{ad}_X|\mathfrak{p} \neq 0$ for $X \neq 0$, compare Exercise B.2. Furthermore, ad_X is skew-symmetric on $\mathfrak{p}$ with respect to the given inner product on $\mathfrak{p}$ since the adjoint representation of K on $\mathfrak{p}$ preserves this product. Now the square of a skew-symmetric endomorphism has negative trace unless the endomorphism vanishes. Therefore B is negative definite on $\mathfrak{k}$. Hence G is semi-simple if and only if the Ricci curvature of $M = G/K$ is non-degenerate.

B.29 Remark (Curvature). Let (G, K) be an infinitesimally effective Riemannian symmetric pair with K/N compact, where N is as in Remark B.27. Since the Ricci tensor of $M = G/K$ is parallel, there are real numbers λ_i and pairwise perpendicular parallel distributions E_i on M such that $\mathrm{Ric}\, X = \lambda_i X$ for all X tangent to E_i. Identify $\mathfrak{p} = T_o M$ and let $X, Y \in \mathfrak{p}$ be tangent to E_i and E_j, respectively. Then, by Theorem B.24.4,

$$\begin{aligned} B([X,Y],[X,Y]) &= -B([X,[X,Y]],Y) = 2\,\mathrm{Ric}([X,[X,Y]],Y) \\ &= 2\lambda_j \langle [X,[X,Y]],Y\rangle = -2\lambda_j \langle R(Y,X)X,Y\rangle, \end{aligned}$$

where we use that Y is tangent to E_j. By the same computation,

$$B([Y,X],[Y,X]) = -2\lambda_i \langle R(X,Y)Y,X\rangle.$$

Hence $B([X,Y],[X,Y]) = 0$ if $\lambda_i \neq \lambda_j$ or if $\lambda_i = 0$ or $\lambda_j = 0$. Since B is negative definite on $\mathfrak{k}$, this implies that the curvature tensor R vanishes on the kernel of Ric. For $\lambda_i \neq 0$ and $X, Y \in \mathfrak{p}$ both tangent to E_i, we get

$$\langle R(X,Y)Y,X\rangle = -\frac{1}{2\lambda_i} B([X,Y],[X,Y]).$$

In particular, the sectional curvature of a symmetric space is non-negative respectively non-positive iff its Ricci curvature is non-negative respectively non-positive.

By definition, M is of compact or, respectively, non-compact type iff all $\lambda_i > 0$ or, respectively, all $\lambda_i < 0$. In particular, a symmetric space of non-compact type is a complete and simply connected Riemannian manifold of non-positive sectional curvature, compare Corollary B.14 and Remark B.15.

If M is simply connected, the distributions E_i give rise to a splitting of M as a Riemannian product with factors M_i tangent to E_i. For each i, M_i is a symmetric Einstein space with Einstein constant λ_i. Applying this to the universal covering space of M, we see that we can renormalize the metric of M without destroying the geometry of M so that the constants λ_i are in $\{\pm 1, 0\}$. For example, if M is of compact type, we can renormalize the metric so that M becomes an Einstein space with Einstein constant 1.

B.30 Remark (Curvature Operator). Let (G, K) be a Riemannian symmetric pair. Suppose that the inner product on $\mathfrak{p}$ is the restriction of an Ad_G-invariant bilinear form[24], also denoted $\langle\cdot,\cdot\rangle$, on $\mathfrak{g}$. Then all ad_X, $X \in \mathfrak{g}$, are skew-symmetric with

[24] For example, a negative or positive multiple of the Killing form of G if (G, K) is infinitesimally effective and $M = G/K$ is of compact or non-compact type, respectively.

respect to $\langle \cdot, \cdot \rangle$, and then, with respect to the identification $\mathfrak{p} = T_o M$,

$$\langle R(X,Y)V, U \rangle = -\langle [[X,Y],V],U \rangle = \langle [X,Y],[U,V] \rangle. \tag{B.31}$$

Recall the definition of the curvature operator $\hat{R}$ in (1.43) and define a morphism

$$F : \Lambda^2 \mathfrak{p} \to \mathfrak{k}, \quad F(X \wedge Y) := [X,Y]. \tag{B.32}$$

By (B.31), we have

$$\langle \hat{R}(X \wedge Y), U \wedge V \rangle = \langle F(X \wedge Y), F(U \wedge V) \rangle. \tag{B.33}$$

In particular, if $\langle \cdot, \cdot \rangle$ is an inner product on $\mathfrak{g}$, then

$$\langle R(X,Y)Y, X \rangle = |[X,Y]|^2 \quad \text{and} \quad \hat{R} = F^* F \geq 0. \tag{B.34}$$

Compare Remarks 1.42 and 5.56 for important consequences of $\hat{R} \geq 0$.

B.35 Remark (De Rham Decomposition). Let M be a simply connected symmetric space. If N is a factor in the de Rham decomposition of M, then N is tangent to one of the distributions E_i as in Remark B.29 (but TN might be smaller than E_i). From Remark B.29 we conclude that N is an Einstein space. Let λ be the Einstein constant of N. We have:

1) If $\lambda < 0$, then N is a symmetric space of non-compact type.
2) If $\lambda = 0$, then N is a Euclidean space.
3) If $\lambda > 0$, then M is of compact type.

In particular, a compact symmetric space is of compact type if and only if its fundamental group is finite.

B.36 Remark (Holonomy). Let (G, K) be a Riemannian symmetric pair and set $M = G/K$. Then $\pi : G \to M$ is a principal bundle with structure group K. The tangent bundle TM is (canonically isomorphic to) the bundle $G \times_K \mathfrak{p}$ associated to the adjoint representation of K on $\mathfrak{p}$. The left-invariant distribution on G determined by $\mathfrak{p}$ is a principal connection which induces the Levi-Civita connection on M. In particular, the holonomy group $\mathrm{Hol}(M)$ of M at $o = [K]$ is contained in the image of K under the adjoint representation on $\mathfrak{p}$.

Assume now that (G, K) is infinitesimally effective with K/N compact, where N is as in Remark B.27. Suppose furthermore that the Ricci curvature of M is non-degenerate. By (B.20), $\mathfrak{h} = [\mathfrak{p}, \mathfrak{p}] + \mathfrak{p}$ is an ideal in $\mathfrak{g}$. By Remark B.29, $B|\mathfrak{k}$ is negative definite. By assumption, $B|\mathfrak{p}$ is non-degenerate. Moreover, $\mathfrak{k}$ and $\mathfrak{p}$ are perpendicular with respect to B, by (B.20). In particular, G is semi-simple and $B|\mathfrak{h}$ is non-degenerate with $\mathfrak{p} \subset \mathfrak{h}$. Hence the B-perpendicular complement $\mathfrak{h}^\perp$ of $\mathfrak{h}$ is contained in $\mathfrak{k}$ and $\mathfrak{h} \cap \mathfrak{h}^\perp = 0$. Since $\mathfrak{h}$ is an ideal in $\mathfrak{g}$, $\mathfrak{h}^\perp$ is an ideal as well. Hence $[\mathfrak{h}, \mathfrak{h}^\perp] = 0$. We conclude that for $X \in \mathfrak{h}^\perp$,

$$e^{tX} e^Y e^{-tX} = e^Y$$

for all $Y \in \mathfrak{p}$ and $t \in \mathbb{R}$. Since $\exp(tX) \in K$, this implies that $\exp(tX)$ acts as the identity on M. Hence $X = 0$ since (G, K) is infinitesimally effective. We conclude that $\mathfrak{g} = \mathfrak{h}$, that is $\mathfrak{k} = [\mathfrak{p}, \mathfrak{p}]$.

For $X, Y \in \mathfrak{p}$, $[X, Y] \in \mathfrak{k}$ and the curvature tensor is $R(X, Y) = -\operatorname{ad}_{[X,Y]}$ on $\mathfrak{p}$. Moreover, the endomorphisms $R(X, Y)$ are in the Lie algebra of the holonomy group $\operatorname{Hol}(M)$. Hence, under the above assumptions on (G, K), the adjoint image of the component K_0 of the identity of K is equal to the reduced holonomy group of M. In particular, if (G, K) is effective and M is of compact or non-compact type, then K_0 is equal to the reduced holonomy group of M. If, moreover, M is simply connected and G is connected, then K is equal to the holonomy group of M at the preferred origin and hence G is equal to the component of the identity of the isometry group of M. Then parallel tensors on M are also G-invariant, compare Theorem B.24.3 for the converse assertion.

B.37 Exercise (Dual Pairs). Let (G, K) be a Riemannian symmetric pair and $\mathfrak{g} = \mathfrak{k} + \mathfrak{p}$ be the associated decomposition of the Lie algebra $\mathfrak{g}$ of G. Show that

$$[X, Y]' := \begin{cases} [X, Y] & \text{if } X \in \mathfrak{k}, \\ -[X, Y] & \text{if } X, Y \in \mathfrak{p} \end{cases}$$

defines a (new) Lie bracket on $\mathfrak{g}$. Note that this change corresponds to passing from $\mathfrak{g}$ to $\mathfrak{g}' = \mathfrak{k} + i\mathfrak{p}$ in the complexification $\mathfrak{g}_{\mathbb{C}}$ of $\mathfrak{g}$. We say that a Riemannian symmetric pair (G', K') is *dual* to (G, K) if the corresponding decomposition of the Lie algebra $\mathfrak{g}'$ of G' is up to isomorphism given by $\mathfrak{g}' = \mathfrak{k} + \mathfrak{p}$ with Lie bracket $[\cdot, \cdot]'$. We also say that the corresponding symmetric spaces are dual. Show:

1) Under duality, the curvature tensor changes sign and compact type corresponds to non-compact type.

2) For each symmetric space M, there is a simply connected symmetric space M' dual to M and M' is unique up to isometry.

3) If M and N are simply connected dual symmetric spaces and K respectively L denotes the group of isometries fixing a point $p \in M$ respectively $q \in N$, then there is an orthogonal transformation $A\colon T_pM \to T_qN$ with $A^*R^N_q = -R^M_p$ and, for any such transformation, $L = AKA^{-1}$.

B.2 Examples. This subsection is devoted to examples which are relevant in our discussion. To keep the presentation transparent, we divide them into three classes, Examples B.38, B.42, and B.53.

B.38 Example. Let $G = \operatorname{Sl}(n, \mathbb{R})$ with involution $\sigma(A) := (A^t)^{-1}$. The set of fixed points of σ is $K = \operatorname{SO}(n)$. The map

$$G/K \to \mathbb{R}^{n^2}, \quad AK \mapsto AA^t,$$

identifies G/K with the space of positive definite symmetric $(n \times n)$-matrices with determinant 1. We view these as the space of normalized inner products on $\mathbb{R}^n$, or, more geometrically via their unit balls, as the space of ellipsoids in $\mathbb{R}^n$ with volume equal to the volume of the Euclidean unit ball. We have

$$\begin{aligned}
\mathfrak{g} &= \mathfrak{sl}(n,\mathbb{R}) = \{A \in \mathbb{R}^{n^2} \mid \operatorname{tr} A = 0\},\\
\mathfrak{k} &= \mathfrak{so}(n) = \{U \in \mathfrak{sl}(n,\mathbb{R}) \mid U^t = -U\},\\
\mathfrak{p} &= \{X \in \mathfrak{sl}(n,\mathbb{R}) \mid X^t = X\}.
\end{aligned}$$

The symmetric bilinear form

$$\langle A, B\rangle = \operatorname{tr} AB \tag{B.39}$$

on $\mathbb{R}^{n^2}$ is invariant under the adjoint action of $\mathrm{Gl}(n,\mathbb{R})$. Its restriction to $\mathfrak{p}$ is positive definite and turns (G, K) into a Riemannian symmetric pair. Since $\langle\cdot,\cdot\rangle$ is Ad_G-invariant, each ad_A, $A \in \mathfrak{g}$, is skew-symmetric with respect to $\langle\cdot,\cdot\rangle$. Therefore we have, for $X, Y, Z \in \mathfrak{p}$,

$$\langle R(X,Y)Y, X\rangle = \langle [X,Y], [X,Y]\rangle \le 0, \tag{B.40}$$

where we note that $\langle\cdot,\cdot\rangle$ is negative definite on $\mathfrak{k}$.

The complexification is $\mathfrak{g}_{\mathbb{C}} = \mathfrak{sl}(n,\mathbb{C}) = \{A \in \mathbb{C}^{n^2} \mid \operatorname{tr} A = 0\}$. We see that

$$\mathfrak{k} + i\mathfrak{p} = \mathfrak{su}(n) = \{U \in \mathfrak{sl}(n,\mathbb{C}) \mid \bar{U}^t = -U\}.$$

Hence $(\mathrm{SU}(n), \mathrm{SO}(n))$ is a dual pair to $(\mathrm{Sl}(n,\mathbb{R}), \mathrm{SO}(n))$, where the involution on $\mathrm{SU}(n)$ is given by the passage to the conjugate matrix. See also Exercise B.59.

B.41 Exercises. 1) Discuss $(\mathrm{Gl}(n,\mathbb{R}), \mathrm{O}(n))$ as a symmetric pair.

2) Show that $\mathrm{Sl}(2,\mathbb{R})/\mathrm{SO}(2)$ with the above Riemannian metric has constant negative curvature -2 and construct isometries to some (other) models of the hyperbolic plane (with curvature -2).

B.42 Example. We show that the Grassmann manifold $G_{\mathbb{F}}(p,q)$ of p-planes in $\mathbb{F}^{p+q}$ together with a natural Riemannian metric is a Riemannian symmetric space, where $\mathbb{F} = \mathbb{R}, \mathbb{C}$, or $\mathbb{H}$. We also discuss the dual symmetric space $G_{\mathbb{F}}^-(p,q)$. These symmetric spaces are also discussed in Examples 2.2.4, 4.10.5, 4.10.6, and B.83.1.

Let $G = \mathrm{O}(p+q)$, $\mathrm{U}(p+q)$, and $\mathrm{Sp}(p+q)$ for $\mathbb{F} = \mathbb{R}, \mathbb{C}$, and $\mathbb{H}$, respectively[25]. The natural action of G on $G_{\mathbb{F}}(p,q)$ is transitive[26]. The stabilizer of the p-plane $\mathbb{F}^p \times \{0\} \subset \mathbb{F}^{p+q}$ is $K = \mathrm{O}(p)\times\mathrm{O}(q)$, $\mathrm{U}(p)\times\mathrm{U}(q)$, and $\mathrm{Sp}(p)\times\mathrm{Sp}(q)$, respectively, hence $G_{\mathbb{F}}(p,q) = G/K$.

Let S be the reflection of $\mathbb{F}^{p+q}$ about $\mathbb{F}^p \times \{0\}$. Then conjugation with S is an involutive automorphism of G with K as its set of fixed points. Hence (G, K) is a symmetric pair.

[25] For the definition of $\mathrm{Sp}(n)$, see the discussion after (B.69).

[26] We multiply vectors in $\mathbb{H}^{p+q}$ with scalars from the right and with matrices from the left.

We write square $(p+q)$-matrices as blocks of four matrices corresponding to the preferred decomposition $\mathbb{F}^{p+q} = \mathbb{F}^p \times \mathbb{F}^q$. In this notation

$$\mathfrak{k} = \left\{ \begin{pmatrix} U & 0 \\ 0 & V \end{pmatrix} \mid \bar{U}^t = -U,\ \bar{V}^t = -V \right\}, \tag{B.43}$$

$$\mathfrak{p} = \left\{ \begin{pmatrix} 0 & -\bar{X}^t \\ X & 0 \end{pmatrix} \mid X \in \mathbb{F}^{q\times p} \right\} \cong \mathbb{F}^{q\times p}. \tag{B.44}$$

We write

$$\begin{pmatrix} A & 0 \\ 0 & B \end{pmatrix} =: D(A,B) \quad \text{and} \quad \begin{pmatrix} 0 & -\bar{X}^t \\ X & 0 \end{pmatrix} =: P(X). \tag{B.45}$$

For $D(A,B) \in K$, $D(U,V) \in \mathfrak{k}$, and $P(X), P(Y) \in \mathfrak{p}$, we have

$$\begin{aligned} \mathrm{Ad}_{D(A,B)} P(X) &= P(BX\bar{A}^t), \\ [D(U,V), P(X)] &= P(VX - XU), \\ [P(X), P(Y)] &= D(\bar{Y}^t X - \bar{X}^t Y, Y\bar{X}^t - X\bar{Y}^t). \end{aligned}$$

There is an Ad_G-invariant inner product on $\mathfrak{g}$,

$$\langle A, B \rangle = \frac{1}{2} \operatorname{Re} \operatorname{tr}(\bar{A}^t B). \tag{B.46}$$

Its restriction to $\mathfrak{p}$ turns (G,K) into a Riemannian symmetric pair. With respect to the identification P of $\mathbb{F}^{q\times p}$ with $\mathfrak{p}$ above, the curvature tensor of $G_{\mathbb{F}}(p,q)$ is given by

$$R(X,Y)Z = X\bar{Y}^t Z + Z\bar{Y}^t X - Y\bar{X}^t Z - Z\bar{X}^t Y. \tag{B.47}$$

By the Ad_G-invariance of the inner product on $\mathfrak{g}$,

$$\langle R(X,Y)Y, X \rangle = |[X,Y]|^2 \geq 0. \tag{B.48}$$

Hence the sectional curvature of $G_{\mathbb{F}}(p,q)$ is non-negative.

Inside $G_{\mathbb{F}}(p,q)$, we consider the open subset $G_{\mathbb{F}}^-(p,q)$ of p-planes on which the non-degenerate form

$$Q_{p,q}(x,y) = -\sum_{j\leq p} \bar{x}_j y_j + \sum_{j>p} \bar{x}_j y_j \tag{B.49}$$

on $\mathbb{F}^{p+q}$ is negative definite. Let $G^- = \mathrm{O}(p,q)$, $\mathrm{U}(p,q)$, and $\mathrm{Sp}(p,q)$, respectively, be the group of linear transformations of $\mathbb{F}^{p+q}$ preserving this form. Then G^- is transitive on $G_{\mathbb{F}}^-(p,q)$ and $G_{\mathbb{F}}^-(p,q) = G^-/K$, where K is as above.

Conjugation with the reflection S of $\mathbb{F}^{p+q}$ about $\mathbb{F}^p \times \{0\}$ is an involutive automorphism of G^- which has K as its set of fixed points. Hence (G^-, K) is a symmetric pair as well. Now we have

$$\mathfrak{p} = \left\{ \begin{pmatrix} 0 & \bar{X}^t \\ X & 0 \end{pmatrix} \mid X \in \mathbb{F}^{q\times p} \right\} \cong \mathbb{F}^{q\times p}, \tag{B.50}$$

whereas K and $\mathfrak{k}$ are as above. We write

$$\begin{pmatrix} 0 & \overline{X}^t \\ X & 0 \end{pmatrix} = P^-(X).$$

Then

$$\begin{aligned} [D(U,V), P^-(X)] &= P^-(VX - XU), \\ [P^-(X), P^-(Y)] &= D(\overline{X}^t Y - \overline{Y}^t X, X\overline{Y}^t - Y\overline{X}^t). \end{aligned}$$

We see that with respect to the identifications P and P^- of $\mathfrak{p}$ with $\mathbb{F}^{q\times p}$ above, the Lie bracket $[\mathfrak{k}, \mathfrak{p}]$ remains the same, but the Lie bracket $[\mathfrak{p}, \mathfrak{p}]$ changes sign. That is, $G_{\mathbb{F}}(p,q)$ and $G_{\mathbb{F}}^-(p,q)$ are dual symmetric spaces in the sense of Exercise B.37. There is a corresponding change in sign for curvature tensor and sectional curvature.

B.51 Exercise. Show that the Riemannian metrics on $G_{\mathbb{F}}(p,q)$ and $G_{\mathbb{F}}^-(p,q)$ are Einsteinian with Einstein constant $k(p+q+2)-4$ and $-k(p+q+2)+4$, respectively, where $k = \dim_{\mathbb{R}} \mathbb{F}$.

B.52 Exercises. 1) Find other representations of Grassmannians as homogeneous spaces, $G_{\mathbb{F}}(p,q) = G/K$, and discuss which of the corresponding pairs (G, K) are Riemannian symmetric pairs.

2) Replacing orthogonal groups by special orthogonal groups, we obtain the Grassmann manifold $G_{\mathbb{R}}^o(p,q)$ of oriented p-planes in $\mathbb{R}^{p+q}$ as homogeneous space G/K, where $G = \mathrm{SO}(p+q)$ and $K = \mathrm{SO}(p) \times \mathrm{SO}(q)$. The same involution σ as above, namely conjugation with the reflection S, leaves G invariant; with respect to it $\mathfrak{k}$ and $\mathfrak{p}$ remain the same as before. Keeping the formula for the inner product on $\mathfrak{p}$, the formula for the curvature tensor remains the same. The natural "forget the orientation" map $G_{\mathbb{R}}^o(p,q) \to G_{\mathbb{R}}(p,q)$ is a twofold Riemannian covering. If $p+q \geq 3$, then $G_{\mathbb{R}}^o(p,q)$ is simply connected.

3) A complex subspace of dimension p in $\mathbb{C}^n$ is also a real subspace of $\mathbb{R}^{2n} = \mathbb{C}^n$ of dimension $2p$, and similarly with $\mathbb{H}^n$. Discuss the corresponding inclusions of Grassmann manifolds.

B.53 Example. Let V be a vector space over a field F. A *symplectic form* on V is an alternating two-form ω on V such that for all non-zero $v \in V$ there is a vector $w \in V$ with $\omega(v,w) \neq 0$. A *symplectic vector space* is a vector space together with a symplectic form. The standard example is F^{2n} with the form

$$\omega((x,y),(u,v)) = \sum (x_\mu v_\mu - y_\mu u_\mu).$$

As in this example, the dimension of a finite-dimensional symplectic vector space V is even, $\dim V = 2n$, and V has a basis $(e_1, \dots, e_n, f_1, \dots, f_n)$ such that

$$\omega(e_\mu, e_\nu) = \omega(f_\mu, f_\nu) = 0 \quad \text{and} \quad \omega(e_\mu, f_\nu) = \delta_{\mu\nu},$$

a *symplectic basis* of V. The choice of a symplectic basis identifies V with the standard symplectic vector space F^{2n}.

Let V be a symplectic vector space over a field F with symplectic form ω. An endomorphism $A\colon V \to V$ is called *symplectic* if the pull back $A^*\omega = \omega$. Since ω is non-degenerate, symplectic endomorphisms are invertible, hence automorphisms. Under composition, the set of all symplectic automorphisms of V is a group, denoted $\mathrm{Sp}(V)$ and, respectively, $\mathrm{Sp}(n, F)$ if $V = F^{2n}$. Let

$$J = \begin{pmatrix} 0 & 1 \\ -1 & 0 \end{pmatrix} \in \mathrm{Gl}(2n, F) \tag{B.54}$$

be the fundamental matrix of the standard symplectic form ω in F^{2n}, where here and below scalars represent the corresponding multiple of the unit matrix or block. Then a linear map $g \in \mathrm{Gl}(2n, F)$ is in $\mathrm{Sp}(n, F)$ if and only if $g^t Jg = J$. Writing g as a matrix of $(n \times n)$-blocks,

$$g = \begin{pmatrix} A & B \\ C & D \end{pmatrix}, \tag{B.55}$$

the condition $g^t Jg = J$ is equivalent to the conditions

$$A^t C = C^t A, \quad B^t D = D^t B, \quad \text{and} \quad A^t D - C^t B = 1. \tag{B.56}$$

It follows that the group of $g \in \mathrm{Sp}(n, F)$ with $B = C = 0$ is an embedded general linear group,

$$\mathrm{Gl}(n, F) \to \mathrm{Sp}(n, F), \quad A \mapsto \begin{pmatrix} A & 0 \\ 0 & (A^t)^{-1} \end{pmatrix}. \tag{B.57}$$

This induces the diagonal embedding of $\mathrm{O}(n, F)$ into $O(2n, F)$.

A subspace L of V is *isotropic* if $\omega(v, w) = 0$ for all $v, w \in L$. A subspace of V is called a *Lagrangian subspace* if it is a maximal isotropic subspace. If the dimension of V is $2n$, then Lagrangian subspaces of V have dimension n. If $(e_1, \dots, e_n, f_1, \dots, f_n)$ is a symplectic basis of V, then the subspace L_0 spanned by $e_1, \dots, e_n$ is Lagrangian. We denote by $G(L, V)$ and $G_F(L, n)$ the space of Lagrangian subspaces of V and F^{2n}, respectively.

Let $F = \mathbb{R}$ and identify $\mathbb{R}^{2n} = \mathbb{C}^n$ by writing $(x, y) = x + iy = z$. We have

$$\langle \bar{v}, w \rangle = \mathrm{Re}\langle \bar{v}, w \rangle + i\omega(v, w),$$

where the left hand side is the standard Hermitian form on $\mathbb{C}^n$ and where $\mathrm{Re}\langle \bar{v}, w \rangle$ and $\omega(v, w)$ are the Euclidean inner product and the standard symplectic form on $\mathbb{R}^{2n}$, respectively. The real subspace L_0 spanned by the unit vectors $e_1, \dots, e_n$ in $\mathbb{C}^n$ is Lagrangian. If L is any other Lagrangian subspace of V and $(b_1, \dots, b_n)$ is an orthonormal basis of L, then $(b_1, \dots, b_n)$ is a unitary basis of $\mathbb{C}^n$ with its standard Hermitian form $\langle \bar{v}, w \rangle$. Therefore the natural action of $\mathrm{U}(n)$ on $\mathbb{C}^n$ induces a transitive action on $G_{\mathbb{R}}(L, n)$ with stabilizer $\mathrm{O}(n)$ of L_0, hence

$$G_{\mathbb{R}}(L, n) = \mathrm{U}(n)/\,\mathrm{O}(n). \tag{B.58}$$

Let $S\colon \mathbb{C}^n \to \mathbb{C}^n$ be the real automorphism $S(z) = \bar{z}$ corresponding to the reflection $S(x, y) = (x, -y)$ of $\mathbb{R}^{2n}$. Then we have $\sigma(A) := SAS^{-1} = \bar{A}$ for any $A \in \mathrm{U}(n)$. Hence the involution σ on $\mathrm{U}(n)$ has $\mathrm{O}(n)$ as its set of fixed points, turning $(\mathrm{U}(n), \mathrm{O}(n))$ into a Riemannian symmetric pair. Thus $G_{\mathbb{R}}(L, n)$ is a symmetric space of dimension $n(n+1)/2$ and a totally geodesic submanifold of the Grassmannian $G_{\mathbb{R}}(n, n)$.

B.59 Exercise. We say that a Lagrangian subspace L of $V = \mathbb{C}^n$ (as above) is *special* if the n-form $dz^1 \wedge \cdots \wedge dz^n$ is real valued on L. Show that the above Lagrangian subspace L_0 is special, that the space $G_{\mathbb{R}}(SL, n)$ of all special Lagrangian subspaces of V is given by the submanifold $\mathrm{SU}(n)/\mathrm{SO}(n)$ of $G_{\mathbb{R}}(L, n) = \mathrm{U}(n)/\mathrm{O}(n)$, and that $G_{\mathbb{R}}(SL, n)$ is totally geodesic in $G_{\mathbb{R}}(L, n)$.

Under the above identification $\mathbb{R}^{2n} = \mathbb{C}^n$, $A \in \mathrm{O}(2n)$ is complex linear iff the pull back $A^*\omega = \omega$. In other words,

$$U(n) = \mathrm{O}(2n) \cap \mathrm{Sp}(n, \mathbb{R}). \tag{B.60}$$

Write $g \in \mathrm{U}(n)$ as $g = A + iB$ with real $(n \times n)$-matrices A and B. Then this identification reads

$$\mathrm{U}(n) \ni A + iB \longleftrightarrow \begin{pmatrix} A & -B \\ B & A \end{pmatrix} \in O(2n) \cap \mathrm{Sp}(n, \mathbb{R}), \tag{B.61}$$

where $A^tB = B^tA$ and $A^tA + B^tB = 1$. In particular, $\mathrm{O}(n) = \mathrm{O}(n, \mathbb{R}) \subset \mathrm{U}(n)$ is diagonally embedded in $\mathrm{O}(2n) \cap \mathrm{Sp}(n, \mathbb{R})$, compare with (B.57) above.

The symmetric space $\mathrm{Gl}(n, \mathbb{R})/\mathrm{O}(n)$ is dual to $\mathrm{U}(n)/\mathrm{O}(n) = G_{\mathbb{R}}(L, n)$ in the sense of Exercise B.37. Motivated by the example of the Grassmannians, we would like to see if we can view $\mathrm{Gl}(n, \mathbb{R})/\mathrm{O}(n)$ as the open set $G_{\mathbb{R}}^{-}(L, n) \subset G_{\mathbb{R}}(L, n)$ of Lagrangian subspaces on which the symmetric form $Q_{n,n}$ from (B.49) is negative definite.

To that end, let $\mathrm{O}(n, n)$ be the orthogonal group of $Q_{n,n}$. Clearly $g \in \mathrm{Gl}(2n, \mathbb{R})$ belongs to $\mathrm{O}(n, n)$ iff

$$A^tB = C^tD, \quad A^tA - C^tC = 1, \quad \text{and} \quad D^tD - B^tB = 1, \tag{B.62}$$

where we write g as a matrix of blocks as in (B.55). Let $g \in \mathrm{O}(n, n) \cap \mathrm{Sp}(n, \mathbb{R})$. By (B.56) and (B.62),

$$B = (A^t)^{-1}C^tD = CA^{-1}D. \tag{B.63}$$

Writing $D = AX$, we compute

$$(1 + C^tC)X = A^tAX = A^tD = 1 + C^tB = 1 + C^tCX. \tag{B.64}$$

Hence $X = 1$, and hence $D = A$ and $B = C$. We conclude that the intersection $\mathrm{O}(n, n) \cap \mathrm{Sp}(n, \mathbb{R})$ consists of the matrices

$$\begin{pmatrix} A & B \\ B & A \end{pmatrix} \tag{B.65}$$

with $A^tB = B^tA$ and $A^tA - B^tB = 1$. We also see that the subgroup of matrices in $\mathrm{O}(n,n) \cap \mathrm{Sp}(n,\mathbb{R})$ with $B = 0$ is a diagonally embedded $\mathrm{O}(n)$.

Let $L_0 \in G_{\mathbb{R}}^-(L,n)$ be the Lagrangian subspace spanned by the first n unit vectors. Let $L \in G_{\mathbb{R}}^-(L,n)$ be another Lagrangian subspace. Let $(b_1,\dots,b_n)$ be an orthogonal basis of L, that is, $Q_{n,n}(b_j,b_k) = -\delta_{jk}$. Let A and B be the $(n \times n)$-matrices with columns the first n and the last n coordinates of the vectors b_j, respectively. Since L is Lagrangian and $(b_1,\dots,b_n)$ is an orthogonal basis of L, the corresponding matrix g as in (B.65) belongs to $\mathrm{O}(n,n) \cap \mathrm{Sp}(n,\mathbb{R})$. Since g maps L_0 to L, we conclude that $\mathrm{O}(n,n) \cap \mathrm{Sp}(n,\mathbb{R})$ is transitive on $G_{\mathbb{R}}^-(L,n)$. The stabilizer of L_0 is the subgroup of matrices in $\mathrm{O}(n,n) \cap \mathrm{Sp}(n,\mathbb{R})$ with $B = 0$, hence

$$G_{\mathbb{R}}^-(L,n) = (\mathrm{O}(n,n) \cap \mathrm{Sp}(n,\mathbb{R}))/\,\mathrm{O}(n). \tag{B.66}$$

Let

$$k := \frac{1}{\sqrt{2}}\begin{pmatrix} 1 & 1 \\ -1 & 1 \end{pmatrix} \in \mathrm{O}(2n) \cap \mathrm{Sp}(n,\mathbb{R}). \tag{B.67}$$

We have

$$\begin{aligned} k\,\mathrm{Gl}(n,\mathbb{R})k^{-1} &= \mathrm{O}(n,n) \cap \mathrm{Sp}(n,\mathbb{R}), \\ k\,\mathrm{O}(n)k^{-1} &= \mathrm{O}(n), \end{aligned} \tag{B.68}$$

where we consider $\mathrm{O}(n)$ and $\mathrm{Gl}(n,\mathbb{R})$ as embedded subgroups in $\mathrm{Sp}(n,\mathbb{R})$ as above. We conclude that conjugation with k identifies $\mathrm{Gl}(n,\mathbb{R})/\,\mathrm{O}(n)$ with

$$(\mathrm{O}(n,n) \cap \mathrm{Sp}(n,\mathbb{R}))/\,\mathrm{O}(n) = G_{\mathbb{R}}^-(L,n). \tag{B.69}$$

Let $F = \mathbb{C}$ and identify $\mathbb{C}^{2n} = \mathbb{H}^n$ by writing $z \in \mathbb{H}^n$ as $z = x + jy$ with $x, y \in \mathbb{C}^{2n}$, where we multiply with scalars from the right. We have $\bar z = \bar x - jy$ and $jx = \bar x j$. For $z = x + jy$ and $w = u + jv$ in $\mathbb{H}^n$, we get

$$\begin{aligned} \langle \bar z, w\rangle &= \langle \bar x - jy, u + jv\rangle \\ &= (\langle \bar x, u\rangle + \langle \bar y, v\rangle) + j(\langle x, v\rangle - \langle y, u\rangle) \\ &= ((x,y),(u,v)) + j\omega((x,y),(u,v)), \end{aligned}$$

where the first term on the right corresponds to the standard Hermitian form and ω to the standard symplectic form on $\mathbb{C}^{2n}$. The group of quaternionic $(n \times n)$-matrices preserving the form $\langle \bar z, w\rangle$ on $\mathbb{H}^n$ is called the *symplectic group*, denoted $\mathrm{Sp}(n)$. Considered as linear maps on $\mathbb{C}^{2n}$, elements of $\mathrm{Sp}(n)$ preserve the Hermitian form and the symplectic form on $\mathbb{C}^{2n}$. For any unit quaternion a,

$$\langle \overline{az}, aw\rangle = \langle \bar z \bar a, aw\rangle = \langle \bar z, w\rangle,$$

so that left multiplication by a defines an element of $\mathrm{Sp}(n)$.

The complex subspace spanned by the unit vectors $e_1,\dots,e_n$ in $\mathbb{H}^n \cong \mathbb{C}^{2n}$ is Lagrangian. If $L \subset \mathbb{C}^{2n}$ is any Lagrangian subspace of $\mathbb{C}^{2n}$ and $(b_1,\dots,b_n)$ is a unitary basis of L, then $\langle \bar{b}_\mu, b_\nu \rangle = \delta_{\mu\nu}$. Hence mapping the standard basis $(e_1, \dots, e_n)$ of $\mathbb{H}^n$ to $(b_1, \dots, b_n)$ defines an element of $\mathrm{Sp}(n)$. It follows that

$$G_{\mathbb{C}}(L, n) = \mathrm{Sp}(n)/\,\mathrm{U}(n). \tag{B.70}$$

The automorphism $s(x + jy) = x - jy$ of $\mathbb{H}$ induces a complex automorphism $S\colon \mathbb{H}^n \to \mathbb{H}^n$, defined by the analogous formula $S(x + jy) = x - jy$. We see that S corresponds to the reflection $S(x, y) = (x, -y)$ of $\mathbb{C}^{2n}$. Since s is an automorphism of $\mathbb{H}$, we have $SAS^{-1} = \sigma(A)$, where $\sigma(A)$ is the matrix obtained by applying s to the entries of A. Clearly σ is an involution of $\mathrm{Sp}(n)$ with $\mathrm{U}(n)$ as its set of fixed points, turning $(\mathrm{Sp}(n), \mathrm{U}(n))$ into a Riemannian symmetric pair. Thus $G_{\mathbb{C}}(L, n)$ is a symmetric space of dimension $n(n+1)$ and a totally geodesic submanifold of $G_{\mathbb{C}}(n, n)$. Since the center of $\mathrm{U}(n)$ is not discrete, $G_{\mathbb{C}}(L, n)$ is a Hermitian symmetric space, see Proposition B.86.

Under the identification $\mathbb{C}^{2n} = \mathbb{H}^n$, $A \in \mathrm{U}(2n)$ is quaternion linear if and only if the pull back $A^*\omega = \omega$. In other words,

$$\mathrm{Sp}(n) = \mathrm{U}(2n) \cap \mathrm{Sp}(n, \mathbb{C}). \tag{B.71}$$

Write $g \in \mathrm{Sp}(n)$ as $g = A + jB$ with complex $(n \times n)$-matrices A and B. Then this identification reads

$$\mathrm{Sp}(n) \ni A + jB \longleftrightarrow \begin{pmatrix} A & -\bar{B} \\ B & \bar{A} \end{pmatrix} \in U(2n) \cap \mathrm{Sp}(n, \mathbb{C}), \tag{B.72}$$

where $A^t B = B^t A$ and $\bar{A}^t A + \bar{B}^t B = 1$. In particular, $\mathrm{U}(n) \subset \mathrm{Sp}(n)$ is diagonally embedded in $\mathrm{U}(2n)$ with blocks A and $\bar{A}$ on the diagonal.

Again we want to identify the dual symmetric space $\mathrm{Sp}(n, \mathbb{R})/\,\mathrm{U}(n)$ with the open subset $G_{\mathbb{C}}^-(L, n) \subset G_{\mathbb{C}}(L, n)$ of Lagrangian subspaces on which the Hermitian form $Q_{n,n}$ from (B.49) is negative definite, where $\mathrm{U}(n) = \mathrm{O}(2n) \cap \mathrm{Sp}(n, \mathbb{R})$ as in (B.60).

We proceed as in the real case. First of all we note that $g \in \mathrm{Gl}(2n, \mathbb{C})$ belongs to the unitary group $\mathrm{U}(n, n)$ of $Q_{n,n}$ iff

$$\bar{A}^t B = \bar{C}^t D, \quad \bar{A}^t A - \bar{C}^t C = 1, \quad \text{and} \quad \bar{D}^t D - \bar{B}^t B = 1, \tag{B.73}$$

where we write g as a matrix of blocks as in (B.55). Let $g \in \mathrm{U}(n, n) \cap \mathrm{Sp}(n, \mathbb{C})$. By (B.56) and (B.73),

$$B = (\bar{A}^t)^{-1}\bar{C}^t D = \bar{C}\bar{A}^{-1} D. \tag{B.74}$$

Writing $D = \bar{A}X$, we compute

$$(1 + C^t\bar{C})X = A^t\bar{A}X = A^t D = 1 + C^t\bar{C}X, \tag{B.75}$$

hence $X = 1$, and hence $D = \bar{A}$ and $C = \bar{B}$. We conclude that the intersection $\mathrm{U}(n, n) \cap \mathrm{Sp}(n, \mathbb{C})$ consists of the matrices

$$\begin{pmatrix} A & B \\ \bar{B} & \bar{A} \end{pmatrix} \tag{B.76}$$

with $\bar{A}^t B = \bar{B}^t A$ and $\bar{A}^t A - \bar{B}^t B = 1$. The subgroup of $g \in \mathrm{U}(n,n) \cap \mathrm{Sp}(n,\mathbb{C})$ with $B = 0$ is a diagonally embedded $\mathrm{U}(n)$ as above. As in the real case, we get

$$G_{\mathbb{C}}^{-}(L,n) = (\mathrm{U}(n,n) \cap \mathrm{Sp}(n,\mathbb{C}))/\mathrm{U}(n). \tag{B.77}$$

Let

$$k := \frac{1}{\sqrt{2}} \begin{pmatrix} 1 & i \\ i & 1 \end{pmatrix} \in \mathrm{U}(2n) \cap \mathrm{Sp}(n,\mathbb{C}). \tag{B.78}$$

We have $k^{-1} = \bar{k}$ and hence

$$k\bar{k}^{-1} = k^2 = \begin{pmatrix} 0 & i \\ i & 0 \end{pmatrix}. \tag{B.79}$$

Let $g \in \mathrm{Sp}(n,\mathbb{C})$. Then $k^{-1}gk \in \mathrm{Sp}(n,\mathbb{R})$ iff $k^{-1}gk = \bar{k}^{-1}\bar{g}\bar{k}$, that is, iff $k\bar{k}^{-1}\bar{g} = gk\bar{k}^{-1}$. The latter holds iff $D = \bar{A}$ and $C = \bar{B}$, that is, iff $g \in \mathrm{U}(n,n) \cap \mathrm{Sp}(n,\mathbb{C})$. We conclude that

$$k\,\mathrm{Sp}(n,\mathbb{R})k^{-1} = \mathrm{U}(n,n) \cap \mathrm{Sp}(n,\mathbb{C}). \tag{B.80}$$

We also have

$$k(\mathrm{O}(2n) \cap \mathrm{Sp}(n,\mathbb{R}))k^{-1} = \mathrm{U}(n), \tag{B.81}$$

where we consider $\mathrm{U}(n)$ on the right as an embedded subgroup in $\mathrm{Sp}(n,\mathbb{C})$ as above. We finally conclude that conjugation with k identifies $\mathrm{Sp}(n,\mathbb{R})/\mathrm{U}(n)$ with

$$(U(n,n) \cap \mathrm{Sp}(n,\mathbb{C}))/\mathrm{U}(n) = G_{\mathbb{C}}^{-}(L,n). \tag{B.82}$$

Since the center of $\mathrm{U}(n)$ is not discrete, $G_{\mathbb{C}}^{-}(L,n)$ is a Hermitian symmetric space, see Remark B.87. Moreover, $G_{\mathbb{C}}^{-}(L,n)$ is biholomorphic to the Siegel upper half plane. For more on this, see e.g. [Bu, Section 31] or [Hel, Exercise VIII.B].

B.3 Hermitian Symmetric Spaces. A *Hermitian symmetric space* is a symmetric space M together with a parallel almost complex structure on M which is compatible with the Riemannian metric on M in the sense of (2.33). Since parallel almost complex structures are complex structures, any such structure turns M into a Kähler manifold.

Symmetric spaces of non-compact type are simply connected (Corollary B.14), but there are symmetric spaces of compact type with non-trivial fundamental group. However, Hermitian symmetric spaces of compact type are Fano manifolds and therefore simply connected, by Theorem 6.12[27].

Let (G, K) be a Riemannian symmetric pair. Then by Theorem B.24.3, compatible parallel complex structures on $M = G/K$ correspond to complex structures on $\mathfrak{p}$ which preserve the inner product on $\mathfrak{p}$ and are invariant under the adjoint action of K on $\mathfrak{p}$.

[27]There are more direct proofs for the latter assertion which do not rely on Kobayashi's theorem, see e.g. [Hel, Theorem VIII.4.6].

B.83 Examples. 1) *Complex Grassmannians.* We follow the notation in Example B.42 and consider

$$G_{\mathbb{C}}(p,q) = \mathrm{U}(p+q)/\,\mathrm{U}(p) \times \mathrm{U}(q) = G/K,$$

the Grassmannian of complex p-planes in $\mathbb{C}^{p+q}$, endowed with the Riemannian metric introduced in (B.46). We define a complex structure on $\mathfrak{p}$ by

$$JP(X) := P(iX),$$

where $X \in \mathbb{C}^{q\times p}$ and $P(X) \in \mathfrak{p}$ is defined as in (B.45). For $D(A,B) \in K$ we have $\mathrm{Ad}_{D(A,B)} P(X) = P(BX\bar{A}^t)$, hence J commutes with Ad_K, turning $G_{\mathbb{C}}(p,q)$ into a Hermitian symmetric space.

The curvature tensor of $G_{\mathbb{C}}(p,q)$ is given in (B.48). For $p = 1$, that is, in the case of complex projective space, we get

$$R(X,Y)Y = \langle Y,Y\rangle X - \langle X,Y\rangle Y + 3\langle X,JY\rangle JY.$$

Thus the holomorphic sectional curvature of complex projective space is constant equal to 4 and the range of its sectional curvature is $[1,4]$. We recommend Karcher's article [Kar] for a nice exposition of the elementary geometry of complex projective spaces.

To compare the Riemannian metric on $G_{\mathbb{C}}(p,q)$ here with the one introduced in Example 4.10, we note that both are invariant under the action of $\mathrm{U}(p+q)$. Thus it suffices to compare them at the plane spanned by the first p unit vectors. For $X \in \mathfrak{p}$, we have

$$e^{tX} = \begin{pmatrix} 1 & -t\overline{X}^t \\ tX & 1 \end{pmatrix} + o(t).$$

Comparing with (4.14), we get that the Riemannian metric there is twice the Riemannian metric discussed here.

The discussion in the case of the dual Grassmannians $G^-_{\mathbb{C}}(p,q)$ is similar. The case $p = 1$ is of particular importance, $\mathrm{U}(1,m)/\,\mathrm{U}(1) \times \mathrm{U}(m) =: \mathbb{C}H^m$ is called *complex hyperbolic space*. It is dual to complex projective space. With respect to the given normalization of the Riemannian metric, the range of its sectional curvature is $[-4,-1]$.

2) *Complex quadric.* Consider the Grassmannian manifold $G^o_{\mathbb{R}}(2,n-2)$ of oriented real 2-planes in $\mathbb{R}^n$ as in Exercise B.52.2,

$$G^o_{\mathbb{R}}(2,n-2) = \mathrm{SO}(n)/\,\mathrm{SO}(2) \times \mathrm{SO}(n-2).$$

There is a complex structure J on $\mathfrak{p}$ which is invariant under the adjoint representation of $\mathrm{SO}(2) \times \mathrm{SO}(n-2)$: Identify $\mathbb{R}^{(n-2)\times 2} \cong \mathbb{C}^{n-2}$, that is, think of the two columns of $X \in \mathbb{R}^{(n-2)\times 2}$ as real and imaginary part of a vector in $\mathbb{C}^{n-2}$, and multiply this vector by i. Thus $G^o_{\mathbb{R}}(2,n-2)$ together with the Riemannian metric induced by the inner product on $\mathfrak{p}$ and the parallel complex structure induced by J is a Kähler manifold.

We can think of $G^o_{\mathbb{R}}(2,n-2)$ as the complex quadric

$$Q = \{[z] \in \mathbb{C}P^{n-1} \mid z_1^2 + \cdots + z_n^2 = 0\}. \qquad \text{(B.84)}$$

To that end, extend the action of $\mathrm{SO}(n)$ on $\mathbb{R}^n$ complex linearly to $\mathbb{C}^n$ to realize $\mathrm{SO}(n)$ as a subgroup of $\mathrm{U}(n)$. Let $p_0 = [1, i, 0, \dots, 0] \in Q$. The orbit of p_0 under the induced action of $\mathrm{SO}(n)$ on $\mathbb{C}P^{n-1}$ is equal to Q. The stabilizer of p_0 is $\mathrm{SO}(2) \times \mathrm{SO}(n-2)$, thus $Q = \mathrm{SO}(n)/\mathrm{SO}(2) \times \mathrm{SO}(n-2) = G^o_{\mathbb{R}}(2, n-2)$.

To compare the metric chosen above with the one on $\mathbb{C}P^{n-1}$, we let

$$A = \begin{pmatrix} A_2 & \\ & 1_{n-2} \end{pmatrix} \quad \text{with} \quad A_2 = \frac{1}{\sqrt{2}} \begin{pmatrix} 1 & i \\ i & 1 \end{pmatrix},$$

where 1_r denotes the unit matrix of size $r \times r$. Since A is unitary, A induces an isometry of $\mathbb{C}P^{n-1}$. Moreover, A maps $p_1 = [1, 0, \dots, 0]$ to p_0. For $X \in \mathfrak{p}$, we get

$$e^{tX} p_0 = e^{tX} A p_1 = A(A^{-1} e^{tX} A) p_1.$$

We have

$$A^{-1} e^{tX} A = \begin{pmatrix} 1_2 & -t\bar{A}_2^t \bar{X}^t \\ tXA_2 & 1_{n-2} \end{pmatrix} + o(t).$$

Since the speed of the curve $e^{tX} p_0$ at $t = 0$ is equal to the speed of the curve $(A^{-1} e^{tX} A) p_1$ at $t = 0$, we get that the Riemannian metric induced from $\mathbb{C}P^{n-1}$, normalized as in Example 1 above, is equal to the chosen metric on $G^o_{\mathbb{R}}(2, n-2)$.

Intersecting two-planes with the unit sphere S^{n-1}, we identify $G^o_{\mathbb{R}}(2, n-2)$ with the space of oriented great circles on S^{n-1}. In particular, $G^o_{\mathbb{R}}(2, 1)$ is a two-sphere or, what amounts to the same, a complex projective line. Thus the quadric $Q \subset \mathbb{C}P^2$ in (B.84) is a complex projective line.

B.85 Remark. Let M be a simply connected Hermitian symmetric space with complex structure J, and let p be a point in M. Then J_p is the differential of an isometry j_p of M fixing p, by Corollary B.6 and (4.43). By Exercise B.2, j_p is in the center of the group of holomorphic isometries of M fixing p. Moreover, since J_p is skew-symmetric and orthogonal, the endomorphisms $\cos x \cdot \mathrm{id} + \sin x \cdot J_p$ of T_pM preserve metric and curvature tensor. Hence, by Corollary B.6, j_p belongs to the component of the identity of the stabilizer of p in the group of holomorphic isometries of M.

Recall the definition of infinitesimally effective from Remark B.27.

B.86 Proposition. *Let (G, K) be an infinitesimally effective Riemannian symmetric pair with G and K compact and connected. Assume that G is semi-simple and that the adjoint representation of K on $\mathfrak{p}$ is irreducible. Then $M = G/K$ admits a G-invariant complex structure compatible with the Riemannian metric on M if and only if the center of K is not discrete, that is, if and only if K is not semi-simple. If the latter holds, then M is a Hermitian symmetric space of compact type, hence simply connected. If, moreover, (G, K) is effective, then the center Z_K of K is a circle and J is (any) one of the two elements of order 4 in Z_K.*

Proof. Since G is compact and semi-simple, M is a symmetric space of compact type. Furthermore, $\pi_1(G)$ is finite and $\pi_2(G) = 0$. The final part of the long exact homotopy sequence associated to the fiber bundle $G \to G/K = M$ is

$$0 = \pi_2(G) \to \pi_2(M) \to \pi_1(K) \to \pi_1(G) \to \pi_1(M) \to 0.$$

It follows that $\pi_1(M)$ is finite. Furthermore, $\pi_2(M)$ is infinite if and only if $\pi_1(K)$ is infinite. Since M is compact with finite fundamental group, the universal covering space of M is compact. Hence $\pi_2(M)$ is finitely generated, and hence $H^2(M, \mathbb{R}) \neq 0$ if and only if $\pi_2(M)$ is infinite.

If M is a compact Hermitian symmetric space, then M is Kählerian and hence $H^2(M, \mathbb{R}) \neq 0$. By what we just said, this implies that $\pi_1(K)$ is infinite. Since K is compact and connected, this happens precisely if the center of K is infinite or, equivalently, if the center of K is not discrete.

Vice versa, assume that the center of K is infinite. Since (G, K) is infinitesimally effective and K is compact, the normal subgroup N as in Remark B.27 is finite. Therefore the center of K/N is infinite as well. Hence we may pass to the quotient pair $(G/N, K/N)$, which is also a Riemannian symmetric pair. In conclusion, we may assume that the adjoint representation of K on $\mathfrak{p}$ is faithful.

Let $Z \subset \mathrm{End}(\mathfrak{p})$ be the space of all endomorphisms of $\mathfrak{p}$ commuting with all Ad_k, $k \in K$. Since the adjoint representation of K on $\mathfrak{p}$ is irreducible, Z is an associative division algebra over $\mathbb{R}$, hence $Z \cong \mathbb{R}, \mathbb{C}$, or $\mathbb{H}$.

Since the adjoint representation of K on $\mathfrak{p}$ is faithful, $K \subset \mathrm{End}(\mathfrak{p})$. Hence the center Z_K of K is a subgroup of $Z^\times$. Since K is compact, Z_K is a compact subgroup of $Z^\times$. Hence if Z_K is not finite, $Z \cong \mathbb{C}$ or $Z \cong \mathbb{H}$. Since maximal Abelian subgroups of $\mathbb{H}^\times$ are isomorphic to $\mathbb{C}^\times$, it follows that Z_K is a circle.

If $J \in Z_K$ is one of the two elements of order 4, then $J^2 = -1$ and J is an Ad_K-invariant complex structure on $\mathfrak{p}$, compatible with the inner product on $\mathfrak{p}$ since $Z_K \subset K$. Thus J turns $M = G/K$ into a Hermitian symmetric space of compact type. In particular, M is simply connected. By Remark B.85, there are no other but the above two choices for an invariant complex structure on M. □

B.87 Remark. Let (G, K) and (G', K') be dual symmetric pairs as in Exercise B.37 with $K = K'$. Then a complex structure J on $\mathfrak{p}$ is invariant under the adjoint representation of K and compatible with the inner product on $\mathfrak{p}$ if and only if the corresponding complex structure J' on $\mathfrak{p}'$ is invariant under the adjoint representation of K' and compatible with the inner product on $\mathfrak{p}'$.

For example, let M be a symmetric space of non-compact type, G be the component of the identity in the group of isometries of M and K be the stabilizer of a point in M. Then M is simply connected and given by the symmetric pair (G, K). Let M' be the simply connected dual symmetric space of M, G' be the group of isometries of M' and K' be the stabilizer of a point in M'. Then M' is of compact type and G' is compact and semi-simple. Up to isomorphism, $\mathfrak{p}' = i\mathfrak{p}$ and hence $K = K'$ by Theorem B.5 and Theorem B.24.4. Hence M is Hermitian iff M' is Hermitian iff $K = K'$ is not semi-simple. In other words, the criterion of Proposition B.86 also applies to symmetric spaces of non-compact type.

B.88 Exercises. 1) Interpret $\mathrm{O}(2n)/\mathrm{U}(n)$ as space of complex structures on $\mathbb{R}^{2n}$ which preserve the Euclidean norm and discuss $(\mathrm{O}(2n), \mathrm{U}(n))$ as a Riemannian symmetric pair.

2) Determine the space of all complex structures on $\mathbb{R}^{2n}$ as a homogeneous space G/K. Interpret $\mathrm{Sp}(n,\mathbb{R})/\mathrm{U}(n)$ as a space of complex structures on $\mathbb{R}^{2n}$.

Let $D \subset \mathbb{C}^m$ be a bounded domain. We say that D is *symmetric* if, for all $p \in M$, there is a biholomorphic transformation $s_p\colon D \to D$ such that $s_p(p) = p$ and $ds_p(p) = -\mathrm{id}$. Recall that the Bergmann metric of D is invariant under all biholomorphic transformations of D, see Example 4.10.7. In particular, transformations s_p as above are geodesic reflections. It follows that symmetric bounded domains with the Bergmann metric are Hermitian symmetric spaces. It then also follows that they are Einstein spaces with Einstein constant -1, see Example 4.10.7, and, hence, that they are of non-compact type.

Vice versa, for any Hermitian symmetric space M of non-compact type, there is a symmetric bounded domain D and a biholomorphic map $M \to D$ which is, up to rescaling, an isometry between M and D, equipped with the Bergmann metric, see Theorem VIII.7.1 in [Hel]. We have seen this explicitly[28] in the case of the dual complex Grassmannians $G^-_{r,n}$, see Example 4.10.6.

We say that $D \subset \mathbb{C}^m$ is a *tube domain* if $D = \mathbb{R}^m + i\Omega$, where $\Omega \subset \mathbb{R}^m$ is an open and convex subset such that, for all $y \in \Omega$ and $t > 0$, $ty \in \Omega$ and such that Ω does not contain complete lines. For example, the upper half plane is a tube domain. Each tube domain in $\mathbb{C}^m$ is biholomorphic to a bounded domain in $\mathbb{C}^m$, but the converse does not hold. The article [Ma2] by Matsushima contains a nice introduction into tube domains.

We say that a Hermitian symmetric space is of *tube type* if it is biholomorphic to a tube domain. The hyperbolic plane is of tube type. The dual Lagrangian Grassmannians $G^-_{\mathbb{C}}(L,n) = \mathrm{Sp}(n,\mathbb{R})/\mathrm{U}(n)$ (as in Example B.53) are of tube type, see Exercise VIII.B in [Hel]. See [BIW], [KW], [Wi] for more information on Hermitian symmetric spaces of tube type.

[28] We did not identify the Bergmann metric there. The irreducibility of the isotropy representation for the two metrics under consideration shows, however, that they must be multiples of each other.

Appendix C Remarks on Differential Operators

Let M be a Riemannian manifold. For a vector bundle $E \to M$, we denote by $\mathcal{E}(M, E)$ and $\mathcal{E}_c(M, E)$ the space of smooth sections of E and smooth sections of E with compact support, respectively. We also use the shorthand $\mathcal{E}(E)$ and $\mathcal{E}_c(E)$ for these spaces. If E is Hermitian, we denote by $L^2(E) = L^2(M, E)$ the Hilbert space of square integrable measurable sections σ of E, endowed with the L^2-inner product from (1.2) and corresponding L^2-norm $\|\sigma\|_2$.

Let $E, F \to M$ be Hermitian vector bundles and D be a differential operator of order $m \geq 0$ from (smooth sections of) E to (smooth sections of) F,

$$D\colon \mathcal{E}(M, E) \to \mathcal{E}(M, F). \tag{C.1}$$

We also view D as an unbounded operator on $L^2(M, E)$ with domain $\mathcal{E}_c(M, E)$. The *formal adjoint* D^* of D is a differential operator of order m from F to E and satisfies, by definition,

$$(D\sigma, \tau)_2 = (\sigma, D^*\tau)_2 \tag{C.2}$$

for all $\sigma \in \mathcal{E}_c(M, E)$ and $\tau \in \mathcal{E}_c(M, F)$. We say that D is *formally self-adjoint* or *formally symmetric* if $E = F$ and $D = D^*$.

By the divergence formula 1.9, an equivalent characterization of D^* is that, for all smooth sections σ of E and τ of F, there is a complex vector field $Z = X + iY$ on M with

$$(D\sigma, \tau) = (\sigma, D^*\tau) + \operatorname{div} Z, \tag{C.3}$$

where the divergence $\operatorname{div} Z := \operatorname{div} X + i \operatorname{div} Y$. This is the way we will determine adjoint operators. The discussion in the case of real vector bundles and Riemannian metrics is similar.

C.4 Example. Let E be a Hermitian vector bundle over M and suppose that E is endowed with a Hermitian connection D. We view D as a differential operator,

$$D\colon \mathcal{E}(E) \to \mathcal{E}(T^*_{\mathbb{C}}M \otimes E). \tag{C.5}$$

Note that $F = T^*_{\mathbb{C}}M \otimes E$ inherits a Hermitian metric from the Hermitian metrics on $T^*_{\mathbb{C}}M$ and E and, by the product rule (1.20), a metric connection $\hat{D}$ from $\hat{\nabla}$ and D. Let σ be a section of E and τ be a section of $T^*_{\mathbb{C}}M \otimes E$, that is, a 1-form with values in E. Let $p \in M$ and $(X_1, \dots, X_n)$ be an orthonormal frame of M about p with $\nabla X_j(p) = 0$. Using the latter and that D is metric, we have at p

$$\begin{aligned}(D\sigma, \tau) &= \sum (D_{X_j}\sigma, \tau(X_j)) \\ &= \sum \big(X_j(\sigma, \tau(X_j)) - (\sigma, (\hat{D}_{X_j}\tau)(X_j))\big) \\ &= -\sum (\sigma, (\hat{D}_{X_j}\tau)(X_j)) + \operatorname{div} Z,\end{aligned} \tag{C.6}$$

where Z is the complex vector field defined by the property $(Z, W) = (\sigma, \tau(W))$. We conclude that

$$D^*\tau = -\sum(\hat{D}_{X_j}\tau)(X_j) = -\operatorname{tr}\hat{D}\tau. \tag{C.7}$$

The same formula holds in the Riemannian context. In both cases, Riemannian or Hermitian,

$$D^*D\sigma = -\sum D^2\sigma(X_j, X_j) = -\operatorname{tr} D^2\sigma. \tag{C.8}$$

The *principal symbol* S_D of a differential operator D as above associates to each $\xi \in T^*M$ a homomorphism $S_D(\xi)\colon E_p \to F_p$, where p is the foot point of ξ. By definition,

$$S_D(d\varphi)\sigma = \frac{1}{m!}\sum_{j=0}^{m}(-1)^j\binom{m}{j}\varphi^j D(\varphi^{m-j}\sigma) \tag{C.9}$$

for all smooth functions φ on M and sections σ of E. The principal symbol is homogeneous of order m in ξ, $S_D(t\xi) = t^m S_D(\xi)$. For example, if D has order one, then S_D is tensorial in $\xi = d\varphi$ and

$$S_D(d\varphi)\sigma = D(\varphi\sigma) - \varphi D(\sigma). \tag{C.10}$$

If $E = F$, then the right hand side of (C.9) can be written as an m-fold commutator,

$$S_D(d\varphi)\sigma = \frac{1}{m!}[[\dots[[D, \varphi], \varphi], \dots, \varphi], \varphi]\sigma, \tag{C.11}$$

where φ is viewed as the operator which multiplies sections with φ. Recall also that D is *elliptic* if $S_D(\xi)$ is invertible for all non-zero $\xi \in T^*M$.

C.12 Exercises. 1) Compute S_D in terms of local trivializations of E.

2) Show that $S_{D^*}(\xi) = (-1)^m S_D(\xi)^*$ for all $\xi \in T^*M$.

3) Let D be a connection on E, viewed as a differential operator as in (C.5). Show that $S_D(d\varphi)\sigma = d\varphi \otimes \sigma$.

The *maximal extension* $D_{\max}$ of D is the adjoint operator of D^* in the sense of Hilbert spaces[29]. By definition, $\sigma \in L^2(M, E)$ belongs to the domain $\operatorname{dom} D_{\max}$ of $D_{\max}$ iff there is a section $\sigma' \in L^2(M, E)$ such that

$$(\sigma', \tau)_2 = (\sigma, D^*\tau)_2 \quad \text{for all } \tau \in \mathcal{E}_c(M, F), \tag{C.13}$$

and then $D_{\max}\sigma := \sigma'$. The operator $D_{\max}$ is closed and extends D. In particular, D is closable. Furthermore,

$$L^2(M, E) = \ker D_{\max} + \overline{\operatorname{im} D^*}, \tag{C.14}$$

an L^2-orthogonal decomposition of $L^2(M, E)$, where we note that $\ker D_{\max}$ is a closed subspace of $L^2(M, E)$.

[29] Section III.5 in [Kat] is a good reference for the functional analysis we need.

C.15 Exercise. Let $\sigma \in \mathcal{E}(M, E)$. Then if σ and $D\sigma$ are square integrable, then $\sigma \in \operatorname{dom} D_{\max}$ and $D_{\max}\sigma = D\sigma$.

The smallest closed extension $D_{\min}$ of D is called the *minimal extension* of D. The domain $\operatorname{dom} D_{\min}$ of $D_{\min}$ consists of all $\sigma \in L^2(M, E)$ such that there is a sequence (σ_n) in $\operatorname{dom} D = \mathcal{E}_c(M, E)$ such that $\sigma_n \to \sigma$ in $L^2(M, E)$ and the sequence $(D\sigma_n)$ is convergent in $L^2(M, F)$, and then $D_{\min}\sigma := \lim D\sigma_n$. By definition, the graph of $D_{\min}$ is the closure of the graph of D and

$$D \subset D_{\min} \subset D_{\max}. \tag{C.16}$$

The left inclusion is always strict. By definition, $\operatorname{im} D_{\min} \subset \overline{\operatorname{im} D}$, hence

$$\overline{\operatorname{im} D_{\min}} = \overline{\operatorname{im} D}. \tag{C.17}$$

The minimal extension $D_{\min}$ can also be characterized as the Hilbert space adjoint of the maximal extension $(D^*)_{\max}$.

C.18 Exercises. 1) Show that $\operatorname{dom} D_{\min}$ endowed with the graph norm of D,

$$|\sigma|_D := (|\sigma|_2^2 + |D_{\min}\sigma|_2^2)^{1/2},$$

is a Hilbert space, the *Sobolev space* associated to D.

2) Assuming $D = D^*$, show that

$$D^a = D_{\max}, \quad D^{aa} = D_{\min}, \quad D^{aaa} = D_{\max},$$

where the superscript a denotes the Hilbert space adjoint.

We say that a differential operator D is *essentially self-adjoint* if it is formally self-adjoint and $D_{\min} = D_{\max}$.

C.19 Theorem (Chernoff [Chf]). *Let D be elliptic, formally self-adjoint, and of order one with S_D uniformly bounded. If M is complete and connected, then any power D^k, $k \geq 1$, of D is essentially self-adjoint.*

The elegant proof of Theorem C.19 in [Chf] uses an existence and uniqueness result for the associated wave equation. In [LM, Theorem II.5.7], the case $k = 1$ is treated by a cut-off argument and by using an appropriate parametrix for D.

C.20 Exercise. Let M be complete and connected and D be elliptic of order one with S_D uniformly bounded. By passing to the elliptic and formally self-adjoint operator

$$\begin{pmatrix} 0 & D^* \\ D & 0 \end{pmatrix},$$

conclude that $A_{\min} = A_{\max}$, where A is any product with alternating factors D and D^*.

C.21 Proposition. *Let D be elliptic of order one with S_D uniformly bounded. If M is complete and connected and $\sigma \in \operatorname{dom}(D^*D)_{\max}$, then $\sigma \in \operatorname{dom} D_{\max}$ and*

$$(D^*D\sigma, \sigma)_2 = |D\sigma|_2^2.$$

Proof. By Exercise C.20, $D_{\max} = D_{\min}$ and $(D^*D)_{\max} = (D^*D)_{\min}$. Let (σ_j) be a sequence of smooth sections of E with compact support such that (σ_j) and $((D^*D)\sigma_j)$ are Cauchy sequences in $L^2(M, E)$. We have

$$\begin{aligned}|D\sigma_j - D\sigma_k|_2^2 &= ((D^*D)(\sigma_j - \sigma_k), (\sigma_j - \sigma_k))_2 \\ &\le |(D^*D)(\sigma_j - \sigma_k)|_2 |(\sigma_j - \sigma_k)|_2,\end{aligned}$$

hence $(D\sigma_j)$ is a Cauchy sequence as well, and hence $\operatorname{dom}(D^*D)_{\min} \subset \operatorname{dom} D_{\min}$. Clearly the asserted formula holds for any σ_j, hence also in the limit. □

C.22 Corollary. *Let D be elliptic of order one with S_D uniformly bounded, and let $\sigma \in \mathcal{E}(M, E)$ be square integrable. If M is complete and connected, then*

$$D^*D\sigma = 0 \iff D\sigma = 0.$$

C.1 Dirac Operators. Among the elliptic differential operators of order one, Dirac operators in the sense of Gromov and Lawson [GL], [LM] play a central role. We say that a Hermitian vector bundle E over a Riemannian manifold M with Hermitian connection ∇ is a *Dirac bundle* if it is endowed with a field of morphisms

$$TM \otimes E \to E, \quad X \otimes \sigma \mapsto X\sigma = X \cdot \sigma,$$

called *Clifford multiplication*, such that

$$\begin{aligned}XY\sigma &= -YX\sigma - 2\langle X, Y\rangle\sigma, \\ (X\sigma, \tau) &= -(\sigma, X\tau), \\ \nabla_X(Y\sigma) &= (\nabla_X Y)\sigma + Y(\nabla_X\sigma).\end{aligned} \tag{C.23}$$

The associated Dirac operator $D\colon \mathcal{E}(E) \to \mathcal{E}(E)$ is then defined by

$$D\sigma = \sum X_j \nabla_{X_j}\sigma, \tag{C.24}$$

where (X_j) is a local orthonormal frame of M. We leave it as an exercise to show that D is formally self-adjoint with principal symbol

$$S_D(d\varphi)\sigma = \operatorname{grad}\varphi \cdot \sigma. \tag{C.25}$$

We see that S_D is uniformly bounded. Hence if M is complete and connected, then any power D^k, $k \ge 1$, of D is essentially self-adjoint, by Theorem C.19. We also note that

$$X \cdot (X \cdot \sigma) = -|X|^2\sigma,$$

hence D is elliptic.

C.26 Exercise. Let E be a Dirac bundle over M. Let F be a Hermitian vector bundle over M, endowed with a Hermitian connection, also denoted ∇. Then $E \otimes F$ with the induced connection and Clifford multiplication $X \cdot (\sigma \otimes \tau) := (X \cdot \sigma) \otimes \tau$ is also a Dirac bundle, the *twist* of E by F. Compare the associated Dirac operators of E and F with the twist construction in (8.12).

Let E be a Dirac bundle and $\mu \colon E \to E$ be a parallel unitary involution anticommuting with Clifford multiplication, $\mu(X\sigma) = -X\mu(\sigma)$. Then the associated eigenbundles $E^\pm$ for the eigenvalues ± 1 of μ are parallel and turn E into a *graded Dirac bundle*, $E = E^+ \oplus E^-$. Furthermore, D restricts to operators $D^\pm \colon \mathcal{E}(E^\pm) \to \mathcal{E}(E^\mp)$. We leave it as an exercise to show that $(D^+)^* = D^-$.

C.27 Examples. 1) The morphism $TM \otimes A^*(M, E) \to A^*(M, E)$,

$$X \cdot \alpha := X^\flat \wedge \alpha - X \lrcorner \alpha,$$

satisfies the axioms for Clifford multiplication in (C.23), see (1.10) and (1.11), and $d + d^*$ is the associated Dirac operator, see (1.21) and Proposition 1.27. We note that, as a Dirac bundle, $A^*(M, E)$ is the twist of $A^*(M, \mathbb{C})$ with E in the sense of Exercise C.26. The splitting $A^*(M, E) = A^{\text{even}}(M, E) \oplus A^{\text{odd}}(M, E)$ turns $A^*(M, E)$ into a graded Dirac bundle.

2) Let M be a complex manifold with complex structure J, endowed with a compatible Riemannian metric g. Let $E \to M$ be a holomorphic vector bundle with Hermitian metric h and associated Chern connection D. For $X \in TM$, let

$$Z_X := \frac{1}{2}(X - iJX) \quad \text{and} \quad Z_X^\flat := \langle Z_X, \cdot \rangle \in A^{0,1}(M, \mathbb{C}),$$

compare Subsection 3.3. Then the morphism $TM \otimes A^{p,*}(M, E) \to A^{p,*}(M, E)$,

$$X \cdot \alpha := \sqrt{2} Z_X^\flat \wedge \alpha - \sqrt{2} \bar{Z}_X \lrcorner \alpha \tag{C.28}$$

satisfies the axioms for Clifford multiplication in (C.23).

Let $(X_1, Y_1, \dots, X_m, Y_m)$ be a local orthonormal frame of M with $JX_j = Y_j$. By (3.32) and (3.33),

$$\begin{aligned} \bar\partial &= \sum \big(Z_{X_j}^\flat \wedge \hat{D}_{X_j} + Z_{Y_j}^\flat \wedge \hat{D}_{Y_j}\big), \\ \bar\partial^* &= -\sum \big(\bar{Z}_{X_j} \lrcorner \hat{D}_{X_j} + \bar{Z}_{Y_j} \lrcorner \hat{D}_{Y_j}\big), \end{aligned}$$

where we use that $iZ_X^\flat = Z_{JX}^\flat$ and $i\bar{Z}_X = -\bar{Z}_{JX}$. It follows that $\sqrt{2}(\bar\partial + \bar\partial^*)$ is the Dirac operator associated to the Clifford multiplication in (C.28).

We note that, as a Dirac bundle, $A^{p,*}(M, E)$ is the twist of $A^{0,*}(M, \mathbb{C})$ with $A^{p,0}(M, E) = A^{p,0}(M, \mathbb{C}) \otimes E$ in the sense of Exercise C.26. As in the previous example, the splitting $A^{p,*}(M, E) = A^{p,\text{even}}(M, E) \oplus A^{p,\text{odd}}(M, E)$ turns $A^{p,*}(M, E)$ into a graded Dirac bundle.

C.2 L^2-de Rham Cohomology. Let $E \to M$ be a Hermitian vector bundle with a flat Hermitian connection D. The exterior differential $d = d^D$ is a differential operator of order one on $A^*(M, E)$. By (C.14), we have L^2-orthogonal decompositions

$$L^2(A^*(M, E)) = \ker d_{\max} + \overline{\operatorname{im} d^*} = \ker d^*_{\max} + \overline{\operatorname{im} d}, \tag{C.29}$$

where $d^*_{\max}$ stands for $(d^*)_{\max}$. Since E is flat, $d^2 = 0$ and hence

$$\operatorname{im} d \subset \operatorname{im} d_{\max} \subset \ker d_{\max}, \quad \operatorname{im} d^* \subset \operatorname{im} d^*_{\max} \subset \ker d^*_{\max}. \tag{C.30}$$

It follows that $\ker d_{\max}$ contains the closure of $\operatorname{im} d_{\max}$, where we recall that $\ker d_{\max}$ and $\ker d^*_{\max}$ are closed subspaces of $L^2(A^*(M, E))$.

We let $\mathcal{H}^*_2(M, E) := \ker d_{\max} \cap \ker d^*_{\max}$ be the space of square integrable *harmonic forms* (with values in E). If α is a harmonic form, then α satisfies $(d + d^*)\alpha = 0$ weakly. Since $d + d^*$ is an elliptic differential operator, see e.g. Exercise C.27.1, it follows that α is smooth and satisfies $(d + d^*)\alpha = 0$ in the classical sense.

C.31 Theorem (Hodge Decomposition). *We have L^2-orthogonal decompositions*

$$\ker d_{\max} = \mathcal{H}^*_2(M, E) + \overline{\operatorname{im} d}, \quad \ker d^*_{\max} = \mathcal{H}^*_2(M, E) + \overline{\operatorname{im} d^*}.$$

Proof. Let $\alpha \in \mathcal{H}^*_2(M, E)$ and $\beta \in \mathcal{A}^*_c(M, E)$. Then $(\alpha, d\beta)_2 = (d^*_{\max}\alpha, \beta)_2 = 0$, hence α is perpendicular to $\operatorname{im} d$. Vice versa, if $\alpha \in \ker d_{\max}$ is L^2-perpendicular to $\operatorname{im} d$, then $d^*_{\max}\alpha = 0$, and hence $\alpha \in \mathcal{H}^*_2(M, E)$. The proof of the second equality is similar. □

By the first inclusion in (C.30), there is a cochain complex

$$\cdots \xrightarrow{d_{\max}} C^{k-1} \xrightarrow{d_{\max}} C^k \xrightarrow{d_{\max}} C^{k+1} \xrightarrow{d_{\max}} \cdots \tag{C.32}$$

with $C^l := \operatorname{dom} d_{\max} \cap L^2(A^l(M, E))$. We let

$$\begin{aligned} Z^k(M, E) &= \ker\{d_{\max} \colon C^k \to L^2(A^{k+1}(M, E))\}, \\ B^k(M, E) &= \operatorname{im}\{d_{\max} \colon C^{k-1} \to L^2(A^k(M, E))\}. \end{aligned} \tag{C.33}$$

The *L^2-de Rham cohomology*[30] of M consists of the quotients

$$H^k_2(M, E) = Z^k(M, E)/B^k(M, E). \tag{C.34}$$

The image $B^k(M, E)$ need not be closed, but its closure $\overline{B}^k(M, E)$ is contained in $\ker d_{\max}$ since the latter is closed in $L^2(A^*(M, E))$. We define the *reduced L^2-de Rham cohomology* of M by

$$H^k_{2,\mathrm{red}}(M, E) := Z^k(M, E)/\overline{B}^k(M, E). \tag{C.35}$$

We note the following consequences of Exercise C.20 and Corollary C.22.

[30] [Car] and [Lot] contain discussions of L^2-de Rham cohomology.

C.36 Proposition. *If M is complete and connected, then the natural projection $\mathcal{H}_2^*(M, E) \to H_{2,\mathrm{red}}^*(M, E)$ is an isomorphism.* □

C.37 Proposition. *Let $\alpha \in \mathcal{A}^*(M, E)$ be square integrable. If M is complete and connected and $\Delta_d \alpha = 0$, then $d\alpha = d^*\alpha = 0$.* □

Suppose now that M is oriented. Since $\bar{*} \otimes h$ commutes with the Laplace operator Δ_d, it induces conjugate linear isomorphisms

$$\mathcal{H}_2^r(M, E) \to \mathcal{H}_2^{n-r}(M, E^*), \tag{C.38}$$

compare (1.56). In particular, Poincaré duality holds for reduced L^2-cohomology if M is oriented, complete, and connected.

C.3 L^2-Dolbeault Cohomology. Let M be a complex manifold with complex structure J, endowed with a compatible Riemannian metric g. Let $E \to M$ be a holomorphic vector bundle with Hermitian metric h and associated Chern connection D. By (C.14), we have L^2-orthogonal decompositions

$$L^2(A^{*,*}(M, E)) = \ker \bar\partial_{\max} + \overline{\operatorname{im} \bar\partial^*} = \ker \bar\partial^*_{\max} + \overline{\operatorname{im} \bar\partial}. \tag{C.39}$$

As in the case of the exterior differential d,

$$\operatorname{im} \bar\partial \subset \operatorname{im} \bar\partial_{\max} \subset \ker \bar\partial_{\max}, \quad \operatorname{im} \bar\partial^* \subset \operatorname{im} \bar\partial^*_{\max} \subset \ker \bar\partial^*_{\max}. \tag{C.40}$$

We let $\mathcal{H}_2^{*,*}(M, E) := \ker \bar\partial_{\max} \cap \ker \bar\partial^*_{\max}$ be the space of square integrable $\bar\partial$-*harmonic forms* (with values in E). If α is a $\bar\partial$-harmonic form, then α satisfies $(\bar\partial + \bar\partial^*)\alpha = 0$ weakly. Since $\bar\partial + \bar\partial^*$ is an elliptic differential operator, see e.g. Exercise C.27.2, it follows that α is smooth and satisfies $(\bar\partial + \bar\partial^*)\alpha = 0$ in the classical sense.

C.41 Theorem (Hodge Decomposition). *We have L^2-orthogonal decompositions*

$$\ker \bar\partial_{\max} = \mathcal{H}_2^{*,*}(M, E) + \overline{\operatorname{im} \bar\partial}, \quad \ker \bar\partial^*_{\max} = \mathcal{H}_2^{*,*}(M, E) + \overline{\operatorname{im} \bar\partial^*}. \qquad \square$$

By the first inclusion in (C.40), there are cochain complexes

$$\cdots \xrightarrow{\bar\partial_{\max}} C^{p,q-1} \xrightarrow{\bar\partial_{\max}} C^{p,q} \xrightarrow{\bar\partial_{\max}} C^{p,q+1} \xrightarrow{\bar\partial_{\max}} \cdots \tag{C.42}$$

with $C^{r,s} := \operatorname{dom} \bar\partial_{\max} \cap L^2(A^{r,s}(M, E))$. We let

$$\begin{aligned} Z^{p,q}(M, E) &= \ker\{\bar\partial_{\max} \colon C^{p,q} \to L^2(A^{p,q+1}(M, E))\}, \\ B^{p,q}(M, E) &= \operatorname{im}\{\bar\partial_{\max} \colon C^{p,q-1} \to L^2(A^{p,q}(M, E))\}. \end{aligned} \tag{C.43}$$

The L^2-*Dolbeault cohomology* of M consists of the quotients

$$H_2^{p,q}(M, E) := Z^{p,q}(M, E)/B^{p,q}(M, E). \tag{C.44}$$

The *reduced L^2-Dolbeault cohomology* of M is given by

$$H^{p,q}_{2,\mathrm{red}}(M,E) := Z^{p,q}(M,E)/\overline{B}^{p,q}(M,E). \tag{C.45}$$

We again note consequences of Exercise C.20 and Corollary C.22.

C.46 Proposition. *If M is complete and connected, then the natural projection $\mathcal{H}^{*,*}_2(M,E) \to H^{*,*}_{2,\mathrm{red}}(M,E)$ is an isomorphism.* □

C.47 Proposition. *Let $\alpha \in \mathcal{A}^{*,*}(M,E)$ be square integrable. If M is complete and connected and $\Delta_{\bar\partial}\alpha = 0$, then $\bar\partial\alpha = \bar\partial^*\alpha = 0$.* □

Since $\bar{*} \otimes h$ commutes with the Laplace operator $\Delta_{\bar\partial}$, it induces conjugate linear isomorphisms

$$\mathcal{H}^{p,q}_2(M,E) \to \mathcal{H}^{m-p,m-q}_2(M,E^*), \tag{C.48}$$

compare (3.11). In particular, Serre duality holds for reduced L^2-Dolbeault cohomology if M is complete and connected.

Suppose now in addition that M is a Kähler manifold and that E is flat. Then $\Delta_d = 2\Delta_{\bar\partial}$. Moreover, the Lefschetz map L commutes with $\Delta_{\bar\partial}$.

C.49 Theorem. *Suppose that M is a complete and connected Kähler manifold and that E is flat. Then the Lefschetz map $L^s\colon \mathcal{H}^k_2(M,E) \to \mathcal{H}^{k+2s}_2(M,E)$ is injective for $0 \le s \le m-k$ and surjective for $s \ge m-k \ge 0$. Furthermore,*

$$\mathcal{H}^r_2(M,E) = \oplus_{p+q=r}\mathcal{H}^{p,q}_2(M,E), \quad \mathcal{H}^{p,q}_2(M,E) = \oplus_{s\ge 0} L^s \mathcal{H}^{p-s,q-s}_{2,P}(M,E),$$

the Hodge and Lefschetz decompositions of $\mathcal{H}^r_2(M,E)$ and $\mathcal{H}^{p,q}_2(M,E)$. □

Literature

[Ad] J. F. Adams, *Lectures on exceptional Lie groups*. With a foreword by J. Peter May, edited by Zafer Mahmud and Mamoru Mimura, Chicago Lectures in Mathematics, University of Chicago Press, Chicago, IL, 1996.

[Ar] V. I. Arnold, *Mathematical methods of classical mechanics*. Translated from the Russian by K. Vogtmann and A. Weinstein, Grad. Texts in Math. 60, Springer-Verlag, New York–Heidelberg, 1978.

[At] M. F. Atiyah, Elliptic operators, discrete groups and von Neumann algebras. *Colloque Analyse et Topologie en l'Honneur de Henri Cartan* (Orsay, 1974); *Astérisque* 32–33 (1976), 43–72.

[Au1] T. Aubin, Métriques riemanniennes et courbure. *J. Differential Geometry* 4 (1970), 383–424.

[Au2] T. Aubin, Équations du type Monge-Ampère sur les variétés kähleriennes compactes. *C. R. Acad. Sci. Paris Sér. A-B* 283 (1976), no. 3, Aiii, A119–A121.

[Au3] T. Aubin, *Nonlinear analysis on manifolds. Monge-Ampère equations*. Grundlehren Math. Wiss. 252, Springer-Verlag, New York, 1982.

[Au4] T. Aubin, Réduction du cas positif de l'équation de Monge-Ampère sur les variétés kählériennes compactes à la démonstration d'une inégalité. *J. Funct. Anal.* 57 (1984), 143–153.

[BT] I. K. Babenko & I. A. Taimanov, On the formality problem for symplectic manifolds. *Global differential geometry, the mathematical legacy of Alfred Gray* (Bilbao, 2000), Contemp. Math. 288, Amer. Math. Soc., Providence, R.I., 2001, 1–9.

[Bea] A. Beauville, Variétés Kähleriennes dont la première classe de Chern est nulle. *J. Differential Geom.* 18 (1983), 755–782 (1984).

[BeG] C. Benson & C. S. Gordon, Kähler and symplectic structures on nilmanifolds. *Topology* 27 (1988), 513–518.

[BGM] M. Berger, P. Gauduchon, & E. Mazet, *Le spectre d'une variété riemannienne*. Lecture Notes in Math. 194, Springer-Verlag, Berlin–New York, 1971.

[Bes] A. L. Besse, *Einstein manifolds*. Ergeb. Math. Grenzgeb. 10, Springer-Verlag, Berlin, 1987.

[Bi] O. Biquard, Métriques kählériennes à courbure scalaire constante. Séminaire Bourbaki, Exposé No. 938, to be published in *Astérisque*.

[Boc] S. Bochner, Vector fields and Ricci curvature. *Bull. Amer. Math. Soc.* 52 (1946), 776–797.

[BM] S. Bochner & D. Montgomery, Groups on analytic manifolds. *Ann. of Math.* 48 (1947), 659–669.

[Bor] A. Borel, *Semisimple groups and Riemannian symmetric spaces*. Texts Read. Math. 16, Hindustan Book Agency, New Delhi, 1998.

[Bo1] J. P. Bourguignon et al., Première classe de Chern et courbure de Ricci: preuve de la conjecture de Calabi. Lectures presented at a Séminaire held at the Centre de Mathématiques de l'École Polytechnique, Palaiseau, March–June 1978, *Astérisque* 58, Société Mathématique de France, Paris, 1978, 169 pp.

[Bo2] J. P. Bourguignon, Eugenio Calabi and Kähler metrics. *Manifolds and geometry* (Pisa, 1993), Sympos. Math. XXXVI, Cambridge University Press, Cambridge, 1996, 61–85.

[Bo3] J. P. Bourguignon, Métriques d'Einstein-Kähler sur les variétés de Fano: obstructions et existence (d'après Y. Matsushima, A. Futaki, S. T. Yau, A. Nadel et G. Tian). Séminaire Bourbaki, Exposé No. 830; *Astérisque* 245 (1997), 277–305.

[Br] R. Bryant, Recent advances in the theory of holonomy. Séminaire Bourbaki, Exposé No. 861; *Astérisque* 266 (2000), 351–374.

[Bu] D. Bump, *Lie groups*. Grad. Texts in Math. 225, Springer-Verlag, New York, 2004.

[BIW] M. Burger, A. Iozzi & A. Wienhard, Hermitian symmetric spaces and Kähler rigidity. Preprint, ETH Zürich, 2005.

[Ca1] E. Calabi, The space of Kähler metrics. *Proceedings of the International Congress of Mathematicians* (Amsterdam, 1954), Vol. 2, Erven P. Noordhoff N.V., Groningen; North-Holland Publishing Co., Amsterdam 1957, 206–207.

[Ca2] E. Calabi, On Kähler manifolds with vanishing canonical class. *Algebraic geometry and topology*, A symposium in honor of S. Lefschetz, Princeton University Press, Princeton, N.J., 1957, 78–89.

[Ca3] E. Calabi, Improper affine hyperspheres of convex type and a generalization of a theorem by K. Jörgens. *Michigan Math. J.* 5 (1958), 105–126.

[CE] E. Calabi & B. Eckmann, A class of compact, complex manifolds which are not algebraic. *Ann. of Math.* 58 (1953), 494–500.

[CX] J. Cao & F. Xavier, Kähler parabolicity and the Euler number of compact manifolds of non-positive sectional curvature. *Math. Ann.* 319 (2001), 483–491.

[CMP] J. Carlson, S. Müller-Stach & C. Peters, *Period mappings and period domains*. Cambridge Stud. Adv. Math. 85, Cambridge University Press, Cambridge, 2003.

[Car] G. Carron, Formes harmoniques L^2 sur les variétés non-compactes. *Rend. Mat. Appl.* (7) 21 (2001), 87–119.

[Che] J. Cheeger, Some examples of manifolds of nonnegative curvature. *J. Differential Geometry* 8 (1973), 623–628.

[CEb] J. Cheeger & D. G. Ebin, *Comparison theorems in Riemannian geometry*. North-Holland Mathematical Library 9, North-Holland Publishing Co., Amsterdam–Oxford; American Elsevier Publishing Co., Inc., New York, 1975.

[CG1] J. Cheeger & D. Gromoll, The splitting theorem for manifolds of nonnegative Ricci curvature. *J. Differential Geometry* 6 (1971/72), 119–128.

[CG2] J. Cheeger & D. Gromoll, On the structure of complete manifolds of nonnegative curvature. *Ann. of Math.* 96 (1972), 413–443.

[CGr] J. Cheeger & M. Gromov, Collapsing Riemannian manifolds while keeping their curvature bounded. I. *J. Differential Geom.* 23 (1986), 309–346.

[Chr] S.-S. Chern, On curvature and characteristic classes of a Riemann manifold. *Abh. Math. Semin. Univ. Hamb.* 20 (1955), 117–126 .

[Chf] P. R. Chernoff, Essential self-adjointness of powers of generators of hyperbolic equations. *J. Functional Analysis* 12 (1973), 401–414.

[De] O. Debarre, Variétés de Fano. Séminaire Bourbaki, Exposé No. 827; *Astérisque* 245 (1997), 197–221.

[DGMS] P. Deligne, P. Griffiths, J. Morgan & D. Sullivan, Real homotopy theory of Kähler manifolds. *Invent. Math.* 29 (1975), 245–274.

[Du] J. J. Duistermaat, *Fourier integral operators*. Progr. Math. 130, Birkhäuser, Boston, MA, 1996.

[DK] J. J. Duistermaat and J. A. C. Kolk, *Lie Groups*. Universitext, Springer-Verlag, Berlin, 2000.

[Eb] P. Eberlein, *Geometry of nonpositively curved manifolds*. Chicago Lectures in Mathematics, University of Chicago Press, Chicago, IL, 1996.

[Eh] P. Ehrlich, Metric deformations of curvature. I. Local convex deformations. *Geom. Dedicata* 5 (1976), 1–23.

[FG] M. Fernández & A. Gray, The Iwasawa manifold. *Differential geometry* (Peñíscola, 1985), Lecture Notes in Math. 1209, Springer-Verlag, Berlin, 1986, 157–159.

[FH] W. Fulton & J. Harris, *Representation theory. A first course*. Grad. Texts in Math. 129, Springer-Verlag, New York, 1991.

[Fu] A. Futaki, An obstruction to the existence of Kähler-Einstein metrics. *Invent. Math.* 73 (1983), 437–443.

[GM] S. Gallot & D. Meyer, Opérateur de courbure et laplacien des formes différentielles d'une variété riemannienne. *J. Math. Pures Appl.* 54 (1975), 259–284.

[GH] P. Griffiths & J. Harris, *Principles of algebraic geometry*. Pure Appl. Math., Wiley-Interscience, New York, 1978.

[Gr] M. Gromov, Kähler hyperbolicity and L^2-Hodge theory. *J. Differential Geom.* 33 (1991), 263–292.

[GL] M. Gromov & H.B. Lawson Jr., Positive scalar curvature and the Dirac operator on complete Riemannian manifolds. *Inst. Hautes Études Sci. Publ. Math.* 58 (1983), 83–196.

[Ha] K. Hasegawa, Minimal models of nilmanifolds. *Proc. Amer. Math. Soc.* 106 (1989), 65–71.

[Hei] H.-J. Hein, *The Ricci flow on Kähler manifolds*. Diplomarbeit, Rheinische Friedrich-Wilhelms-Universität Bonn, Bonn, 2006.

[Hel] S. Helgason, *Differential Geometry and Symmetric Spaces*. Pure Appl. Math. XII, Academic Press, New York–London, 1962.

[Hir] F. Hirzebruch, *Topological methods in algebraic geometry*. Appendix one by R. L. E. Schwarzenberger, appendix two by A. Borel, reprint of the 2nd, corr. print. of the 3rd ed. 1978, Classics Math., Springer-Verlag, Berlin, 1995.

[Hit] N. Hitchin, On the curvature of rational surfaces. *Differential geometry* (Stanford Univ., Stanford, Calif., 1973), Part 2, Proc. Sympos. Pure Math. XXVII, Amer. Math. Soc., Providence, R.I., 1975, 65–80.

[Hu] D. Huybrechts, *Complex geometry. An introduction*. Universitext, Springer-Verlag, Berlin, 2005.

[JZ] J. Jost & K. Zuo, Vanishing theorems for L^2-cohomology on infinite coverings of compact Kähler manifolds and applications in algebraic geometry. *Comm. Anal. Geom.* 8 (2000), 1–30.

[Jo] D. D. Joyce, *Compact manifolds with special holonomy*. Oxford Math. Monographs, Oxford University Press, Oxford, 2000.

[Kä1] E. Kähler, Über eine bemerkenswerte Hermitesche Metrik. *Abh. Math. Semin. Hamb. Univ.* 9 (1932), 173-186.

[Kä2] E. Kähler, *Mathematische Werke. Mathematical works*. Edited by Rolf Berndt and Oswald Riemenschneider, Walter de Gruyter, Berlin, 2003.

[Kar] H. Karcher, Riemannian comparison constructions. *Global differential geometry*, MAA Stud. Math. 27, Math. Assoc. America, Washington, DC, 1989, 170–222.

[Kat] T. Kato, *Perturbation theory for linear operators*. Reprint of the 1980 edition, Classics Math., Springer-Verlag, Berlin, 1995.

[Ka1] J. Kazdan, A remark on the preceding paper of S. T. Yau, "On the Ricci curvature of a compact Kähler manifold and the complex Monge-Ampère equation. I". *Comm. Pure Appl. Math.* 31 (1978), 413–414.

[Ka2] J. Kazdan, *Applications of Partial Differential Equations to Problems in Geometry*. Lecture Notes, University of Pennsylvania, Philadelphia, Pa, 2004.

[Ko1] S. Kobayashi, On compact Kähler manifolds with positive definite Ricci tensor. *Ann. of Math.* 74 (1961), 570–574.

[Ko2] S. Kobayashi, *Differential geometry of complex vector bundles*. Publ. Math. Soc. Japan 15, Kanô Memorial Lectures 5, Princeton University Press, Princeton, N.J.; Iwanami Shoten, Tokyo, 1987.

[KN] S. Kobayashi & K. Nomizu, *Foundations of differential geometry*. Vol. II, reprint of the 1963 original, Wiley Classics Lib., John Wiley & Sons, Inc., New York, 1996.

[KW] A. Korányi & J. A. Wolf, Realization of hermitian symmetric spaces as generalized half-planes. *Ann. of Math.* 81 (1965), 265–288.

[LM] H. B. Lawson & M.-L. Michelsohn, *Spin geometry*. Princeton Math. Ser. 38, Princeton University Press, Princeton, N.J., 1989.

[Li1] A. Lichnerowicz, Transformation infinitésimales conformes de certaines variétés riemanniennes compactes. *C. R. Acad. Sci. Paris* 241 (1955), 726–729.

[Li2] A. Lichnerowicz, Sur les transformations analytiques des variétés kählériennes compactes. *C. R. Acad. Sci. Paris* 244 (1957), 3011–3013.

[Li3] A. Lichnerowicz, *Géométrie des groupes de transformations*. Travaux et Recherches Mathématiques III, Dunod, Paris, 1958.

[Loh] J. Lohkamp, Curvature h-principles. *Ann. of Math.* 142 (1995), 457–498.

[Lot] J. Lott, L^2-cohomology of geometrically infinite hyperbolic 3-manifolds. *Geom. Funct. Anal.* 7 (1997), 81–119.

[Ma1] Y. Matsushima, Sur la structure du groupe d'homéomorphismes analytiques d'une certaine variété kählérienne. *Nagoya Math. J.* 11 (1957), 145–150.

[Ma2] Y. Matsushima, On tube domains. *Symmetric spaces* (Short Courses, Washington Univ., St. Louis, Mo., 1969–1970), Pure and Appl. Math. 8, Dekker, New York, 1972, 255–270.

[MS] J. Milnor & J. Stasheff, *Characteristic classes*. Ann. of Math. Stud. 76, Princeton University Press, Princeton, N.J.; University of Tokyo Press, Tokyo, 1974.

[Mo] A. Moroianu, *Kähler geometry*. Lecture Notes 2004.

[Na] A. M. Nadel, Multiplier ideal sheaves and Kähler-Einstein metrics of positive scalar curvature. *Ann. of Math.* 132 (1990), 549–596.

[NN] A. Newlander & L. Nirenberg, Complex analytic coordinates in almost complex manifolds. *Ann. of Math.* 65 (1957), 391–404.

[Ob] M. Obata, Certain conditions for a Riemannian manifold to be isometric with a sphere. *J. Math. Soc. Japan* 14 (1962), 333–340.

[PP] D. N. Page & C. N. Pope, Inhomogeneous Einstein metrics on complex line bundles. *Classical Quantum Grav.* 4 (1987), 213–225.

[Ro] J. Roe, *Elliptic operators, topology and asymptotic methods*. Pitman Res. Notes Math. Ser. 179, Longman S&T, Harlow, 1988.

[Sa] T. Sakai, *Riemannian geometry*. Translated from the 1992 Japanese original by the author. Transl. Math. Monogr. 149, Amer. Math. Soc., Providence, R.I., 1996.

[ShS] B. Shiffman & A. Sommese, *Vanishing theorems on complex manifolds*. Progr. Math. 56, Birkhäuser, Boston, MA, 1985.

[Si1] Y. T. Siu, *Lectures on Hermitian-Einstein metrics for stable bundles and Kähler-Einstein metrics*. DMV Seminar 8, Birkhäuser Verlag, Basel, 1987.

[Si2] Y. T. Siu, The existence of Kähler-Einstein metrics on manifolds with positive anti-canonical line bundle and a suitable finite symmetry group. *Ann. of Math.* 127 (1988), 585–627.

[Th] R. P. Thomas, Notes on GIT and symplectic reduction for bundles and varieties. Preprint, arXiv:math.AG/0512411.

[Ti1] G. Tian, On Kähler-Einstein metrics on certain Kähler manifolds with $C_1(M) > 0$. *Invent. Math.* 89 (1987), 225–246.

[Ti2] G. Tian, On Calabi's conjecture for complex surfaces with positive first Chern class. *Invent. Math.* 101 (1990), 101–172.

[Ti3] G. Tian, *Canonical metrics in Kähler geometry*. Notes taken by Meike Akveld, Lectures Math. ETH Zürich, Birkhäuser Verlag, Basel, 2000.

[TY] G. Tian & S. T. Yau, Kähler-Einstein metrics on complex surfaces with $C_1 > 0$. *Comm. Math. Phys.* 112 (1987), 175–203.

[TO] A. Tralle & J. Oprea, *Symplectic manifolds with no Kähler structure*. Lecture Notes in Math. 1661, Springer-Verlag, Berlin, 1997.

[Va] R. Vakil, *Complex algebraic surfaces*. Lecture Notes, Stanford University, 2002.

[Wa] F. W. Warner, *Foundations of differentiable manifolds and Lie groups*. Scott, Foresman and Co., Glenview, Ill.-London, 1971.

[Wei] A. Weil, *Introduction à l'étude des variétés kählériennes*. Publications de l'Institut de Math. de l'Université de Nancago, VI. Actualités Sci. Ind. no. 1267, Hermann, Paris 1958.

[Wel] R. O. Wells, *Differential analysis on complex manifolds*. Prentice-Hall Series in Modern Analysis, Prentice-Hall, Inc., Englewood Cliffs, N.J., 1973.

[Wi] A. K. Wienhard, *Bounded cohomology and geometry*. Dissertation, Rheinische Friedrich-Wilhelms-Universität Bonn, Bonn 2004, Bonner Mathematische Schriften 368, Universität Bonn, Mathematisches Institut, Bonn, 2004.

[Wo] J. A. Wolf, *Spaces of constant curvature*. Fourth edition, Publish or Perish, Inc., Houston, TX, 1977.

[Ya] K. Yano, On harmonic and Killing vector fields. *Ann. of Math.* 55 (1952), 38–45.

[Ya1] S. T. Yau, On the curvature of compact Hermitian manifolds. *Invent. Math.* 25 (1974), 213–239.

[Ya2] S. T. Yau, Calabi's conjecture and some new results in algebraic geometry. *Proc. Nat. Acad. Sci. U.S.A.* 74 (1977), 1798–1799.

[Ya3] S. T. Yau, On the Ricci curvature of a compact Kähler manifold and the complex Monge-Ampère equation. I. *Comm. Pure Appl. Math.* 31 (1978), 339–411.

[Zh] F. Zheng, *Complex differential geometry*. AMS/IP Stud. Adv. Math. 18, Amer. Math. Soc., Providence, R.I.; International Press, Boston, MA, 2000.

Index